Electronics and Telecommunications Research

Electronics and Telecommunications Research

Microstrip Antennas Modeling for Recent Applications
Amel Boufrioua
2016. ISBN: 978-1-63485-251-7 (Hardcover)
2016. ISBN 978-1-63485-535-8 (eBook)

Smart Miniaturized Wideband/Multiband and Reconfigurable Antenna for Modern Applications
Amel Boufrioua
2018. ISBN: 978-1-53612-816-1 (Softcover)
2018. ISBN 978-1-53613-151-2 (eBook)

Telecommunications: Applications, Modern Technologies and Economic Impact
Joseph P. Barringer (Editor)
2014. ISBN: 978-1-63117-141-3 (Hardcover)
2014. ISBN 978-1-63117-142-0 (eBook)

More information about this series can be found at
https://novapublishers.com/product-category/series/Electronics and Telecommunications Research /

S. Kannadhasan and R. Nagarajan

Recent Trends in Microstrip Antennas for Wireless Applications

Copyright © 2022 by Nova Science Publishers, Inc.
DOI: https://doi.org/10.52305/WFIA7227

All rights reserved. No part of this book may be reproduced, stored in a retrieval system or transmitted in any form or by any means: electronic, electrostatic, magnetic, tape, mechanical photocopying, recording or otherwise without the written permission of the Publisher.

We have partnered with Copyright Clearance Center to make it easy for you to obtain permissions to reuse content from this publication. Simply navigate to this publication's page on Nova's website and locate the "Get Permission" button below the title description. This button is linked directly to the title's permission page on copyright.com. Alternatively, you can visit copyright.com and search by title, ISBN, or ISSN.

For further questions about using the service on copyright.com, please contact:
Copyright Clearance Center
Phone: +1-(978) 750-8400   Fax: +1-(978) 750-4470   E-mail: info@copyright.com.

NOTICE TO THE READER

The Publisher has taken reasonable care in the preparation of this book, but makes no expressed or implied warranty of any kind and assumes no responsibility for any errors or omissions. No liability is assumed for incidental or consequential damages in connection with or arising out of information contained in this book. The Publisher shall not be liable for any special, consequential, or exemplary damages resulting, in whole or in part, from the readers' use of, or reliance upon, this material. Any parts of this book based on government reports are so indicated and copyright is claimed for those parts to the extent applicable to compilations of such works.

Independent verification should be sought for any data, advice or recommendations contained in this book. In addition, no responsibility is assumed by the Publisher for any injury and/or damage to persons or property arising from any methods, products, instructions, ideas or otherwise contained in this publication.

This publication is designed to provide accurate and authoritative information with regard to the subject matter covered herein. It is sold with the clear understanding that the Publisher is not engaged in rendering legal or any other professional services. If legal or any other expert assistance is required, the services of a competent person should be sought. FROM A DECLARATION OF PARTICIPANTS JOINTLY ADOPTED BY A COMMITTEE OF THE AMERICAN BAR ASSOCIATION AND A COMMITTEE OF PUBLISHERS.

Additional color graphics may be available in the e-book version of this book.

Library of Congress Cataloging-in-Publication Data

ISBN: 978-1-68507-744-0

Published by Nova Science Publishers, Inc. † New York

Contents

Preface		vii
Chapter 1	Overview of Microstrip Patch Antennas	1
Chapter 2	Bandwidth Enhancement Techniques of Microstrip Patch Antennas in Modern Communication	13
Chapter 3	Microstrip Patch Antennas Using Various Structures for Various Applications	33
Chapter 4	Recent Trends in Metamaterials, Challenges and Opportunities	53
Chapter 5	5G and 4G Communication for Microstrip Patch Antennas	63
Chapter 6	Wearable Antennas for Microstrip Patch Antennas	77
References		85
About the Authors		91
Index		93

Preface

The book covers a broad range of topics, including basic antenna theory, analytical and numerical techniques in applied electromagnetics, antenna arrays (including adaptive), aperture antennas, antenna measurements, microwave engineering, industrial and medical microwave applications, and so on. 5G propagation, MIMO and array antennas, Optical nano-antennas, Scattering and diffraction, Computational electromagnetics, Radar systems, Plasmonics and nanophotonics, and Advanced EM materials and structures such as metamaterials and metasurfaces are among the subjects covered in the book.

Chapter 1

Overview of Microstrip Patch Antennas

Abstract

This chapter provides a literature analysis of the WLAN and WiMAX implementations for microstrip patch antenna. In this rapidly evolving world of wireless communication, dual or multiband antenna has played a crucial role in the demands of wireless service. Antenna is essentially a transitional guide and is used to radiate or absorb radio waves. Microstrip patch antenna has many advantages such as low cost, compact size, easy structure and integrated circuit compatibility. This has major applications in military, radar, mobile communications, global positioning systems (GPS), remote sensing and so on. Wireless local area network (WLAN) and Worldwide Interoperability for Microwave Access (WiMAX) have been widely applied in mobile devices such as handheld computers and smart phones.

1. Introduction

Wireless networking is the fastest-growing technology sector which has captured social life's attention in the present century. For several households, business canters and campuses new wireless local area networks are introduced. Microstrip antennas have attractive characteristics of low profile, light weight, simple manufacturing and mounting host-conformance. A Microstrip unit simply means a two-layer parallel conductive sandwich separated by a single thin dielectric substratum. The lower conductor is called Ground Plane, and the upper conductor is a plain circular/rectangular resonant patch. The metallic patch (usually Cu or Au) may take many geometrics via. rectangular, circular, triangular, elliptical, helical, ring, etc.

In: Recent Trends in Microstrip Antennas for Wireless Applications
Editors: S. Kannadhasan and R. Nagarajan
ISBN: 978-1-68507-744-0
© 2022 Nova Science Publishers, Inc.

Microstrip patch antennas have variety of feed techniques. Such strategies can be divided into two contacting and non-contacting types. Use a microstrip line as a connecting element the RF power is fed directly to the radiating patch in the contacting process. Electromagnetic field coupling is performed in the non-contacting scheme to transfer power between the microstrip line and the radiating layer. Microstrip antenna feed may have various configurations, such as microstrip line, coaxial, aperture coupling, and proximity coupling. But the microstrip line is comparatively simpler to produce and the coaxial feed. Coaxial probe feed is used because it is user friendly and coaxial cable input impedance in general is 50 ohm. The patch has many points that have an impedance of 50 ohm. A Microstrip antenna (MSA) is well suited for wireless communication due to its low profile, conformal character, light weight, low production cost, robust design and compatibility with Microwave Monolithic Integrated Circuits (MMICs) and OEIC technology.

The antenna engineering industry has an 80-year tradition. The primary part of a wireless communication network is an antenna. Microstrip and printed circuit have gained the maximum attention of the antenna community in recent years with the development of MIC and HF semi-conductor devices. Antenna Microstrip was first introduced in the 1950s. Nevertheless, this idea to wait for the realization of around 20 years after the introduction of printed circuit board (PCB) technology in the 1970s. Microstrip antenna found application in different fields due to its compact size, they have widely engaged for the civilian and military application such as radio- frequency identification (RFID), broadcast radio, mobile system, vehicle collision avoidance system, satellite communications, surveillance system, direction finding, radar systems, remote sensing, and missile guidance and so on. Amid various attractive features the microstrip dimension suffers from an inherent narrow bandwidth disadvantage and low gain. A new pattern or solutions is found in microstrip antenna researcher where the researcher attempts to increase bandwidth by incorporating different structures inside the antenna geometry.

2. Parameters of Microstrip Antennas

Various parameters are measured such as VSWR, Return Loss, Antenna Gain, Directivity, Antenna Capacity and Bandwidth. Gain is the parameter which measures the degree of direction of the radiation pattern of the antenna. The ratio of the radiated power Pr to the input power Pi is defined as. The input power is converted into radiated power and surface wave power while a small portion of the materials used are dissipated due to conductive and dielectric losses.

The radiation pattern is described as a function of space coordinates as a mathematical function, or a graphical representation of the antenna's radiation properties. Antenna Efficiency is a ratio of the total power radiated from an antenna to an antenna's input power. The standing wave tension ratio is known as VSWR = Vmax/Vmin. Around 1 and 2 it would be misleading. In a standing wave pattern, VSWR is defined as the ratio of maximum voltage to minimal voltage. Return loss is a result of signal power from a system being inserted into a transmission line.

The RL is given as by as:

$$RL = -20 \log10 (\Gamma) Db$$

The radiation efficiency is defined as the ratio of power radiated in space to total input power, which is the amount of power dissipated by loss of the conductor, power dissipated by dielectric loss, and radiation from the surface waves. The patch antenna's radiation efficiency is influenced not only by conductor and dielectric losses but also by the excitation of the surface wave. When the thickness of the substratum decreases, the effect of the conductive and dielectric losses becomes more extreme, limiting the output, while the excitation of the surface wave limits the output due to the mutual coupling of elements in an array that causes undesirable diffraction of the edge of the ground plane.

The surface wave phase velocity is highly dependent on the dielectric constant and substrate thickness. The antenna bandwidth is defined as the frequency range, over which the antenna's output complies with a given standard with respect to some characteristic. The impedance bandwidth of the microstrip patch antenna is defined as the impedance variance with the patch antenna element frequency results in a frequency range restriction, over which the device can be matched to its fed line. Impedance bandwidth is generally described in terms of a return loss or a maximum SWR (defined in terms of the coefficient of input reflection) typically less than 2.0 or 1.5 over a frequency range. MSA's radiation bandwidth is defined as the frequency range in which the radiating power is within 3 dB of the incident power and effectively the same radiation pattern. It is the frequency range across which the antenna holds its polarization. The major disadvantage of microstrip patch antenna is narrow bandwidth but higher operating bandwidth is expected today for wireless communication systems. Narrow bandwidth from printed microstrip patches was recognized as one of the most significant factors restricting this type of antenna wide-spread use.

3. Antenna Design

To design a microstrip patch antenna following parameters such as dielectric constant (€r), resonant frequency (fo), and height (h) are considered for calculating the length and the width of the patch.

1. Width Of Patch (W)
2. Effective dielectric constant of antenna (€reff)
3. Effective Length of antenna
4. The extended length of antenna (ΔL)
5. The length of the patch

4. Applications of Microstrip Antennas

There are some benefits of microstrip patch antenna over traditional microwave antenna. The Microstrip patch antennas are well known for their performance, durable design, manufacturing, and extent of use. The advantages of this Microstrip patch antenna are to resolve their de-merits such as simple design, light weight, etc. the applications are widespread in various fields such as medical devices, satellites and, of course, also in military systems such as rockets, aircraft missiles, etc.

Mobile communication includes antennas which are lightweight, low-cost and low profile. Microstrip patch antenna meets all specifications and for use in mobile communication systems different types of microstrip antennas have been developed. Circularly polarized radiation patterns are needed in satellite communication and can be realized either using square or circular patch with one or two feed points. Nowadays microstrip patch antennas are used for global positioning system with substrate having high permittivity sintered material.

Wi-Max is recognized as the IEEE 802.16 standard. This will potentially reach up to 30 mile radius and data rate 70 Mbps. MPA produces three resonant modes at 2,7, 3,3 and 5,3 GHz and can therefore be used in networking equipment which is compatible with WiMax. Radar can be used to track movable objects including people and vehicles. It needs a subsystem with low profile, light weight antennas, the microstrip antennas are an ideal option. The photolithography-based manufacturing technology enables large-scale development of microstrip antenna with repeatable output at lower cost compared to traditional antennas in a lower time frame.

Device antenna in telemedicine operates at 2.45 GHz. Wearable microstrip antenna suits the Wireless Body Area Network (WBAN) framework. In addition to the semi-directional radiation pattern, which is favoured to the omni-directional pattern to mitigate unwanted radiation to the user's body, the antenna obtained a higher gain and back ratio compared to the other antennas, which meets the necessity for on-body and off-body applications. For telemedicine applications an antenna with a gain of 6.7 dB and an F/B ratio of 11.7 dB that resonates at 2.45GHz is ideal.

This is found that the microwave radiation is claimed to be the most effective means of causing hyperthermia in the treatment of malignant tumours. The construction of the particular radiator to be used for this purpose should be compact, simple to handle and rough. Those specifications are fulfilled only by the patch radiator. The original prototypes for the Microstrip radiator to cause hyperthermia were based on the dipoles and annular rings printed on S-band. And later on the design was based at L-band on the circular microstrip disk.

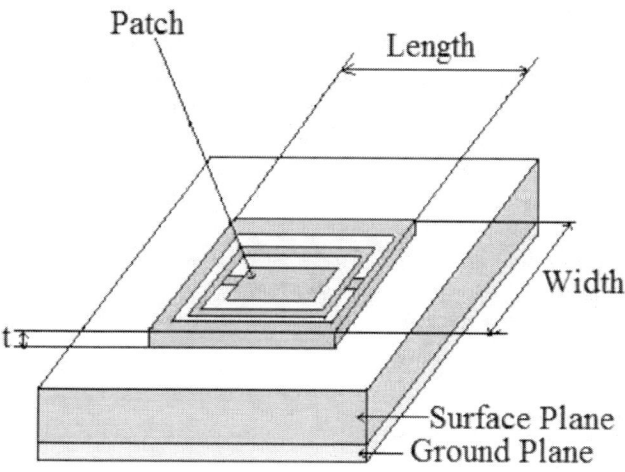

Figure 1.1. Microstrip patch antenna.

5. Feeding Techniques

Coaxial Feed configuration A coaxial connector's inner conductor extends through the substratum and is attached to the radiating patch while the outer conductor is attached to the ground plane. The position of the feed pin is chosen to give the best match for impedance. This system is simple to produce but is suffering from low bandwidth and erroneous radiation. Thicker substrates will elevate the surface wave and produce a high polarized cross. Edge Feed This design attaches a conductive stripe directly to the edge of the microstrip patch. This form of feed arrangement has the advantage of being processed with the picture etch along with the patch itself, thus reducing production.

Aperture Coupled In this arrangement a ground plane is sandwiched, dividing the radiating area, between two layers of the substrate material. Coupling between the patch and the feed line is rendered in the ground plane by means of a slot or aperture, which is normally cantered underneath the patch. Proximity Coupled configuration. The entire antenna consists of a grounded substratum where there is a line of feed for microstrips. There is a further dielectric layer above this material, with a microstrip patch etched on its top surface. The power from the feed network is electromagnetically coupled to the board, thereby providing an alternative to the contacting feed techniques deficiencies. Unlike the more inductive direct contact approaches, the coupling mechanism of the proximity-coupled patch is capacitive in nature. The disparity in coupling greatly influences the impedance bandwidth received, and therefore the bandwidth of a proximity-coupled patch is greater than the direct contact feed patches.

6. Polarization

The simplest definition of polarisation is the direction in which a radio wave's electric field oscillates while propagating across a medium. Looking at the signal from the transmitter's perspective is the point of reference for defining polarisation. Imagine being right behind a radio antenna and gazing in the direction it is pointed to picture this. The electric field will travel sideways on a horizontal plane if the polarisation is horizontal. Vertical polarisation, on the other hand, causes the electric field to oscillate up and down in a vertical plane. An antenna system that uses both horizontal and vertical polarisation is known as linear polarisation. The two polarizations are orthogonal to one another. Orthogonality enables a particular antenna polarisation to exclusively receive on its intended polarisation, avoiding interference from radiation on the orthogonal polarisation. Even though the two orthogonal polarizations are running on the same frequency/channel, this is the case. The polarisation of a slant polarisation antenna is at −45 degrees and +45 degrees from a reference plane of 0 degrees, rather than horizontal and vertical.

Despite the fact that this is simply another kind of linear polarisation, the word linear is often used to refer to just H/V polarisation antennas. Slant polarisation is comparable to turning a linear polarisation radio 45 degrees, using the same analogy as being behind the radio and gazing in the direction of the signal. It is possible to send a signal that seems to spin in polarisation as it travels from the transmitter to the receiver. A circular polarisation is what this is called (CP). Right Hand Circular Polarization (RHCP) and Left Hand Circular Polarization (LHCP) are the two distinct ways in which the signal may spin (LHCP). The rotation of the CP signal in three dimensions is represented as 360°, which should not be mistaken with a complete wavelength. When looking at a 2D side view of phase shifted radio waves, the phase shifting that happens to produce a CP signal is 90°, which equals a quarter wavelength offset, as seen in picture 5. The effect will appear in three-dimensional space as a spinning signal, either in the left or right hand direction, depending on which direction the 90° phase shift happens, i.e., whether H is 90° ahead of V or vice versa. The electrical vector of the wave spins through a complete rotation in a single wavelength, but the two waves stay linear and orthogonal during the transmission.

Microstrip antenna essentially displays two polarization forms that are linear and circular. The linearly polarized patch exhibits electric field variance in one direction only. Based on the patch orientation this polarization can be either vertical or horizontal. The one big trouble with linear polarization is variable or unknown antenna orientation. The circularly polarized patch antenna offers an appealing alternative over linear polarization, provides more versatility in the angle between transmitting and receiving antennas, decreases the effect of multitrack reflections, increases weather penetration and allows transmitter mobility and receiving

Several techniques have been employed in single fed microstrip antenna to achieve circular polarization. Truncate corner of patch antenna is one of the simple perturbation techniques used to achieve circular polarization. Thus circular polarization using such technique can be accomplished by building a patch in orthogonal directions with two resonance frequencies and using the antenna right between the two resonances.

7. Simulation Softwares

HFSS is the industry standard modelling method for simulating full-wave electromagnetic fields in 3D. HFSS includes E- and H-fields, currents, S-parameters, and tests from near and far radiated fields. The automated solution process, in which users are only needed to determine geometry, material properties and the desired performance, is intrinsic to the success of HFSS as an engineering design tool. HFSS will generate an effective, secure, and accurate mesh for solving the problem automatically.

Advanced Design System is the world's leading platform for automation of electronic architecture for wireless RF, microwave, and high speed applications. ADS pioneers the most revolutionary and commercially popular developments, such as X-parameters and 3D EM simulators, used by leading companies in the wireless communication, networking, aerospace & defence industries, in a efficient and easy to use gui. ADS offers complete, standard-based design and testing with Wireless Libraries and circuit-system-EM co-simulation in an integrated platform for WiMAX, LTE, multi-gigabit per second data connections, radar & satellite applications.

CST Microwave Studio (CST MWS) is a specialized device for simulating high frequency components with 3D EM. CST MWS 'has achieved unprecedented success making it the first choice in R&D departments leading in technology. CST MWS allows high frequency (HF) components such as antennas, filters, couplers, planar and multi-layer structures, and SI and EMC effects to be measured rapidly and reliably.

8. Advantages

1. Low Weight
2. Low Profile
3. Thin Profile
4. Required No Cavity Backing
5. Linear and Circular Polarization
6. Dual and Triple Frequency Operation
7. Feedline and Matching Network can be fabricated simultaneously

9. Disadvantages

1. Low Efficiency
2. Low Gain
3. Large Ohmic Loss in the feed structures of array
4. Low Power Handling Capacity
5. Excitation of Surface Waves
6. Polarization Purity is difficult to achieve
7. Complex feed structures requires high performance arrays

Table 1.1. Microstrip patch antennas publication details of conference and journals

Sl.No	Year	Conference	High Quality Journals
1.	2010	150	120
2.	2011	250	180
3.	2012	280	250
4.	2013	350	280
5.	2014	380	320
6.	2015	420	380
7.	2016	580	400
8.	2017	620	480
9.	2018	680	520
10.	2019	700	600
11.	2020	800	620

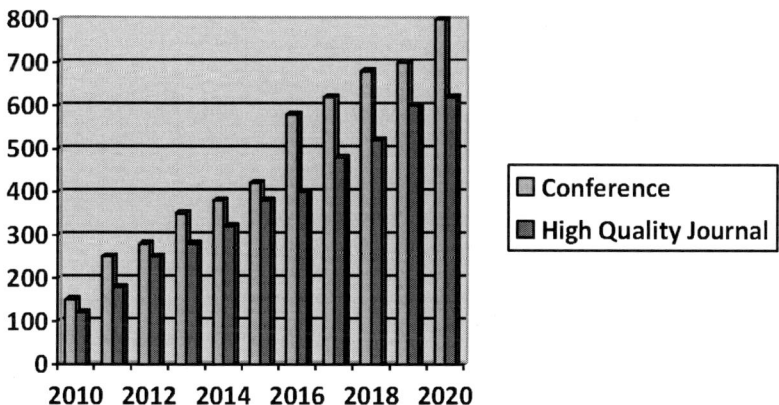

Figure 1.2. Comparison chart for published paper in conference and journal.

Due to its relevance is an active microstrip patch antenna research field for various journal publications in the year of 2010-2020 is shown in Table 1 and its Comparison Chart Publication in 2010–2020 is shown in Figure 2.

Table 1.2. Performance analysis of return loss, gain and VSWR

Sl.No	Frequencies (GHz)	Return Loss (dB)	Gain (dB)	VSWR
1.	2.5	-35.05	9.2	1.8
2.	3.5	-32.08	8.5	1.7
3.	4.5	-30.25	8.3	1.6
4.	5.5	-28.25	7.8	1.8
5.	6.5	-25.45	7.5	1.6
6.	7.5	-22.45	7.2	1.5
7.	8.5	-20.48	8.5	1.4
8.	9.5	-19.05	7.8	1.5
9.	10.5	-18.05	8.2	1.3
10.	11.5	-17.15	6.5	1.2
11.	12.5	-16.20	5.5	1.4

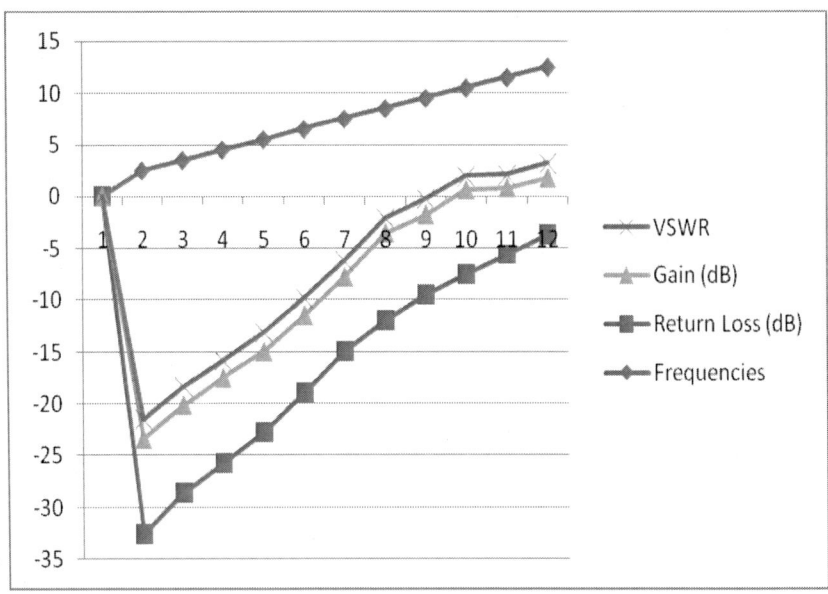

Figure 1.3. Comparison analysis of return loss, gain and VSWR.

Due to its relevance is an active microstrip patch antenna research field for various parameters is shown in Table 2 and its Comparison Chart Publication in 2010–2020 is shown in Figure.3.

10. Conclusion

This chapter examines the technological advancements in the field of microstrip patch antennas. The technological advancement of microstrip antennas is increasing every day. A lot of work is being done on the microstrip antenna for future applications. The majority of methods are created by balancing the Microstrip Antenna strength and bandwidth. According to the survey, few publications on the Microstrip antenna were published in the early years, but then there was a rise in numbers until 2010-2020, after which there was a continuous decrease in the number of articles published. There has been a lot of software developed to simulate microstrip antennas.

Chapter 2

Bandwidth Enhancement Techniques of Microstrip Patch Antennas in Modern Communication

Abstract

Microstrip antennas (MSAs) are used in high-performance aircraft, rockets, satellites, and missiles, where there are limitations on scale, weight, expense, performance, ease of installation, and aerodynamic profile. Square, rectangular, circle, triangle, and every other arrangement can be a radiating patch. Analytical and computational approaches for constructing microstrip patch antennas have been used in the past. Radio signal reception in order to prevent self-interference (SI) induced by simultaneous uplink and downlink frequency transmission and reception. Just half of the throughput of a full-duplex device running at a typical radio frequency (RF) carrier can be accomplished through such duplexing techniques. If the wireless transceiver is allowed to send and receive concurrently at the same RF, which is called full-duplex (IBFD) in-band service, the data throughput of the wireless communication device may be doubled. Potential cellular networks, the biggest problem is to handle heavy data traffic within a small bandwidth.

1. Introduction

The reader transmits RF energy to detect the tag in passive UHF RFID systems, mainly as a transmitter, a tag consists of a processor and an antenna and is powered by energy from the obtained signal, if there is enough power to power up the RFID IC, then the tag reacts to the reader and the information stored in the tag chip is backscattered from the tag to the reader along with data from the reader. In the operation of all radio devices, the

In: Recent Trends in Microstrip Antennas for Wireless Applications
Editors: S. Kannadhasan and R. Nagarajan
ISBN: 978-1-68507-744-0
© 2022 Nova Science Publishers, Inc.

antenna plays a major function. Antennas are important parts of radio technology that are used in the areas of radio transmission, television broadcasting, two-way radio, receivers, radar, wireless telephones, satellite communications, and other equipment. An antenna is an array of conductors (elements) attached to the receiver or transmitter electrically. The oscillating current added to the antenna by a transmitter generates an oscillating electric field and magnetic field across the components of the antenna during transmission. As a travelling transverse electromagnetic field wave, these time-varying fields radiate energy out from the antenna to space. In comparison, the oscillating electric and magnetic forces of the incoming radio wave impose influence on the electrons in the components of the antenna during absorption, allowing them to travel back and forth, producing the antenna's oscillating currents.

Because of its low profile, light weight and simplicity of integration with monolithic integrated microwave circuits (MMIC), slot antennas have received much interest. Numerous methods have been proposed to extend and enhance the efficiency of printed slot antennas' bandwidth. Slot antennas may then be realized by utilizing either a micro strip line or the arrangement of the CPW feed line. A specification of a micro strip line fed with a fork-like tuning stub printed slot antenna for bandwidth enhancement and sufficient radiation has been tested. Although it contributes to a dynamic configuration because of the inclusion of micro strip line feed with a fork-like tuning stub. That would also contribute to the issue of misalignment. If a CPW feed is used to excite the slot, the alignment errors may be avoided. The feed structure of the micro strip line excludes the high frequency resonance, whereas the low frequency part of the UWB spectrum is eliminated by the CPW feed structure.

A further specification for a rectangular slot antenna fed by CPW that has a basic structure and less parameters. A 50W CPW transmission line with a triangular formed tuning stub enabling ultra wideband features is presented in the build. This arrangement increases the impedance spectrum and spans the whole Ultra wideband of frequencies, eliminating the high and low frequency limits of UWB range. Often considered are parametric tests and radiation characteristics. A new rectangular slot antenna fed by a 50W CPW was presented for UWB applications. By utilizing a triangular formed tuning stub, a broad bandwidth of 131 percent was achieved. For the entire UWB frequency spectrum (3.1 GHz to 10.6 GHz), the VSWR impedance bandwidth lower than 2($|S11|$ <-10 dB) was accomplished with a basic antenna structure. The antenna supports the bandwidth and radiation pattern of impedance. It is possible to characterize Radio Frequency Identification (RFID) as follows: automated identification technology that uses electromagnetic radio frequency fields to recognize items bearing tags when they come near to a reader. For several industrial services, the RFID device in the ultra-high frequency (UHF) band is more desirable because it is capable of delivering high reading speed, capable of multiple accesses, long reading distance compared to other RFID frequency band systems, so it has been commonly used in several service sectors, logistics buying and delivery, business, manufacturing firms and materials.

2. RFID Microstrip Patch Antenna

Balanced feed line that is placed in a way to deliver the conjugate impedance matching between the antenna and the chip, the RFID chip is inserted into the patch. For many factors, the miniaturization of RFID UHF tag antennas is of great interest: enabling tiny art facts to be detected, lower prices, less content and faster output. Through a change in the resonant frequency, the I-shaped slot helps to reduce the scale of the tag antenna. To fix this transition, the measurements of the patch may be modified to monitor the resonant frequency to restore it to 915Hz. The antenna has a narrower scale with a dimension of 68.4mm to 75.1mm relative to the other antenna (optimized antenna), and therefore a simplified configuration that guarantees fast and low-cost production. Actually, by deploying two independent half-duplex channels with various frequencies or time slots for transmitting and time, wireless networking systems accomplish maximum duplex service.

IBFD wireless is one of the new innovations for wireless networks of the next decade and is actually being studied for 5G networks as it has the ability to double wireless networking systems' data throughput. Not only are IBFD systems spectrally effective, they are also low-cost since they can conveniently use multiple-input multiple-output transceivers for maximum duplex service. In current wireless networking networks, IBFD may also help address certain issues, such as secret terminals, reduced throughput induced by congestion and major delays in the network. While the IBFD method improves the performance and offers several advantages over conventional wireless full-duplex approaches, there are a range of challenges to IBFD service, including antenna architecture. When the IBFD terminal appears to send and receive concurrently at the same carrier level, the key issue for IBFD wireless service is the synchronization of the transmitting signal to its receiver.

By suppressing the SI at the receiver that is triggered by coupling from its own transmitter, IBFD wireless contact activity can be realized. For IBFD transceivers that share single antennas for transmitting and receiving RF signals, both direct coupling and ambient reflections induce SI. The amount of SI cancellation (SIC) required is based on the transmitted signal strength and bandwidth. Normally, for IBFD service, SI suppression of 100 dB or more is necessary. The SI suppression function is typically applied at three levels through the IBFD transceiver to accomplish this volume of SI suppression, and they are classified as antenna cancellation, RF/analogue cancellation and optical base-band cancellation. In order to relax the necessary amount of SIC at the remaining two stages, much of the SI suppression is done at the antenna level. Transmit and receive chains share a similar antenna in the mutual design of the antenna.

Transmitting and receiving purposes, the IBFD transceivers with a single microstrip patch antenna generate good SI for the interport coupling between transmit and receive antenna ports. Antenna stage SI suppression methods seek to minimize the intrinsic reciprocal coupling between transmit and receive antenna ports by first utilizing transmit and receive activity cross-polarization and then using SIC external active circuitry to gain additional separation. A dual-port dual-polarized microstrip antenna recorded by two separate feeding methods for a single circular patch antenna achieves low cross-polarization and high interport isolation. In the first setup, one port is fed by a probe and another port is coupled by an H-shaped aperture. In the second setup, one port is excited by a differential feeding network that uses a pair of L-shaped probes with a phase difference of 180° and by H-shaped aperture coupling, the other port excitation is achieved. More than 40 dB interport separations is achieved by the implemented antenna.

By altering the ground plane, another means of growing the impedance bandwidth of the microstrip patch antennas may be accomplished. To maximize bandwidth and antenna geometry, novel shapes of changed ground plane are used as trapezoidal form and utilizing proximity feed. Recently, owing to their excellent pass and rejection frequency band properties, electromagnetic band gap (EBG) structures have drawn a great deal of interest among researchers in the microwave and antenna communities. The principle by using electromagnetic structure as 2D-EBG etching as a dumb-bell form on the feed line to boost the antenna's bandwidth and compare it with the prototype antenna's bandwidth for the same feed location, increase the pass band, decrease the size of the antenna and eliminate the harmonic wave. The optimized antenna structure functions from 2 to 35 GHz in the frequency spectrum, which implies that it has an impedance bandwidth of approximately 1000 percent from the basic resonant frequency. A system that may radiate or absorb electromagnetic waves is an antenna. It is possible to identify an antenna as transmitting and receiving. It is also known that the same antenna may be used for electromagnetic wave transmission and reception. The capacity to concentrate and form the radiated power in space is the essential property of an antenna. For successful transmission and reception, the present contact scenario uses microwave frequencies. For radiation at frequencies from 1GHz to 300GHz, the term microwave is used. In the late 1970s, the accelerated advancement of the microstrip antenna system began. In terms of architecture and simulation, simple microstrip antenna elements and arrays were reasonably well known by the early 1980s, and staff shifted their focus to enhancing antenna efficiency characteristics e.g., bandwidth) and growing the implementation of the technology. One of the microwave antenna forms is the microstrip patch antenna (MPA).

The function of microstrip antennas has become enormous in modern telecommunications and radio-electronic systems at the moment. This antenna is found in a broad range of applications because of all of the benefits of these systems is shown in Figure 1. A substance with optimum electrical permeability should be used in order to minimize the stray area of microstrip lines and to decrease its geometric measurements. Its narrow bandwidth is the big issue with the microstrip antenna. The multi-layer microstrip patch antenna structure can be used to achieve broader bandwidth efficiency. A metal conducting patch on the top side consists of a microstrip antenna which can be written on a thin grounded dielectric surface called a substrate. In the simplest situations, power can be supplied by a single radiating part via a coaxial line, a microstrip line or by an electro-magnetic coupling.

3. Printed Circuit Antenna

A three-port microstrip patch antenna uses a 3 dB ring hybrid coupler as a SIC circuit to cancel the direct coupled RF signal from the transmitting port of the antenna (centre port) to transmit ports (side ports) to achieve high interport separation through differential mode excitation for Rx service. Since the antenna is fed for Tx and Rx activity from the same tip, it is orthogonally polarized due to the differential process of excitation for Rx operation. Due to their advantages such as lightweight, small scale, and low production costs, PRINTED circuit antennas such as microstrip or strip line have been common for a few decades. They have some disadvantages, however, such as low performance, low strength, weak purity of polarization, and very small bandwidth; and these drawbacks in the nature of microstrip antennas by traditional methods are very difficult to resolve. By exciting the elliptical patch rather than rectangular or circular ones, circular polarization can be conveniently achieved by careful selection of both the feed location and the ellipse eccentricity, i.e., the feed point should be located on a radial line relative to the main axis where the positive sign induces left-hand circularly polarized polarization. Owing to the advancement of new integrated circuit technologies, the sizes and weight of different wireless electronic devices (e.g., Smartphone handsets) have decreased steadily. There is a need for low profile antennas in certain wireless networking schemes. These antennas are less obstructive and moreover, their output is less influenced by snow, rain or wind.

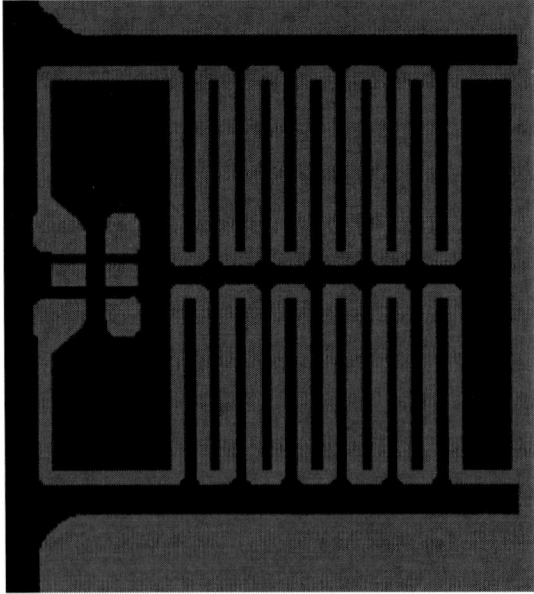

Figure 1. RFID antenna.

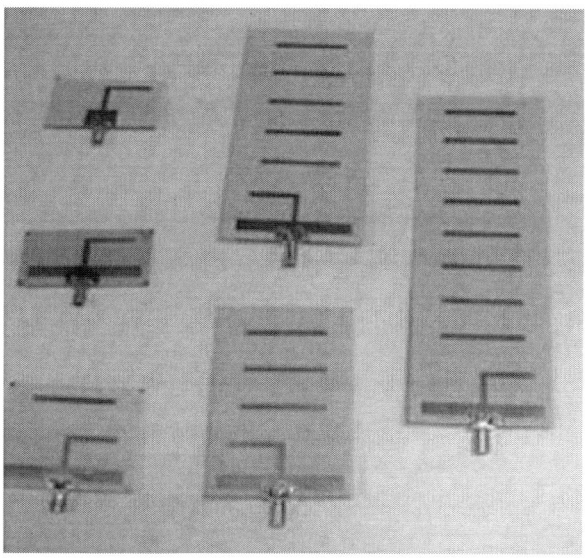

Figure 2. Printed circuit antenna.

Analytical approaches are most useful for functional designs as well as for offering a strong intuitive interpretation of the function of MSAs, focused on certain basic simplification of physical principles about the radiation process of antennas. However for certain structures, these approaches are not sufficient, in particular if the substrate thickness is important. The computational approaches are mathematically complicated and within a rational amount of time, may also not make a realistic antenna configuration feasible. They will need good awareness of the context and have time-consuming computational simulations that involve very pricey software packages. Recently numerous improved approaches used in the design of microstrip patch antennas, including the usage of multiple types of artificial intelligence, have been published in several journals. To efficiently measure and refine the patch measurements of a rectangular microstrip patch antenna fed by a coaxial probe, a procedure focused on the adaptive neuro-fuzzy inference system (ANFIS) is provided. Wireless cell networking networks would be more mature and more common in the future is shown in Figure 2. This rise has created an immense demand for not only power, but also improved coverage and higher service quality. Smart antenna is one of the most promising developments that will allow a higher bandwidth in wireless networks by effectively reducing multipath and co-channel interference. By directing the radiation only in the target direction and adapting itself to shifting traffic patterns or signal environments, this can be accomplished. Adaptive array antennas are also classified as smart antennas. Smart antennas or adaptive clusters are able to respond dynamically to the changing requirements of traffic.

4. Adaptive Antenna

The technique of integrating the signals and then directing the radiation in a certain direction is sometimes referred to as the creation of digital beams. It is primarily used on the mobile target computer to measure beam forming vectors and to map & locate the antenna beam. Higher network bandwidth is the most noticeable one. It improves network capability by correctly monitoring the quality of signal nulls and mitigating distortion coupled with frequency reuse. It raises network provider profits and offers clients less risk of blocked or dropped calls. Adaptive Beam Shaping is a technique in which an arrangement of antennas is used by calculating the signal arrival from the target direction (in the presence of noise) to obtain optimum transmission in a given direction, whereas signals of the same frequency from other directions are refused. A basic and adaptive smart antenna algorithm, the LMS algorithm is commonly used in adaptive due to its relatively low complexity, strong stability characteristics, and relatively good implementation error robustness. The lowest mean square (LMS) algorithm however has sluggish convergence, which decreases the efficiency of the method. The LMS algorithm, which is known as the Recursive Least Mean Square Algorithm (RLMS), is updated recursively in order to maximize the convergence pace. Patch antenna has different advantages such as low profile, light weight, compact volume and integrated microwave circuit (MIC) and monolithic integrated microwave circuit (MMIC) congruent. Nevertheless in large applications for the microstrip antenna, the limited bandwidth is the biggest impediment. The impedance bandwidth of the traditional microstrip antenna is usually just a limited percentage (2% - 5%). While rectangular and circular geometries are primarily used in recent communication systems, where the key interest is compactness, other geometries with greater size reduction find large applications. To operate at two separate frequencies at a distance quite far from each other, methods such as the Global Location System (GPS) and Global System for Mobile Communications (GSM) are required.

The usage of two diverse single band antennas can be prevented by micro-strip antennas. Multiple approaches for achieving dual frequency processing have been suggested. Among them the most commonly seen are loading slits, using slots in the patch, loading the patch with short buttons, using layered patches, or using two feeding ports. Additionally, there are planar antennas with specific geometries to attain dual-band operation. A typical microstrip patch antenna consists primarily of a radiating metallic patch on one side of a dielectric substrate and on the other side of the ground plane. For both constructed frequency bands, typically reconfigurable antennas provide the same radiation patterns and make optimal usage of electromagnetic spectrum and frequency selectivity, which helps to reduce the undesirable impact of co-site disturbance and jamming. Multiband antennas also played a crucial role in wireless service needs through the exponential advancement of wireless connectivity networks, removing the need for different antennas for each application. One of the most common networks for accessing the internet is the wireless local area network (WLAN).

WLAN has become commonly accepted as a feasible, cost-effective and high-speed data networking option that enables device mobility. In WLAN implementations, much focus has been given to the need for multiband operations, such as the IEEE 802.11 b/standards spanning 2.4, 5.2 and 5.8 GHz bands. Using one monopole for the lower band and another monopole or a branch arrangement for the two higher bands is a basic solution used in WLAN antennas to protect the three bands. However this technique contributes to a comparatively broad antenna size due to the necessary duration of the monopole for resonating in the lower band. To excite a dual band resonance, a pair of symmetrical horizontal strips inserted in a slot on the ground plane was used. Technology has raised the market for high gain and wideband operating frequencies of lightweight microstrip antennas. Performance enhancement is expected to accommodate the challenging bandwidth. There are various and well-known approaches to improve antenna bandwidth, including raising substrate thickness, utilizing a low dielectric substrate, slotted patch antenna, using different strategies for balancing impedance and feeding is shown in Figure 3.

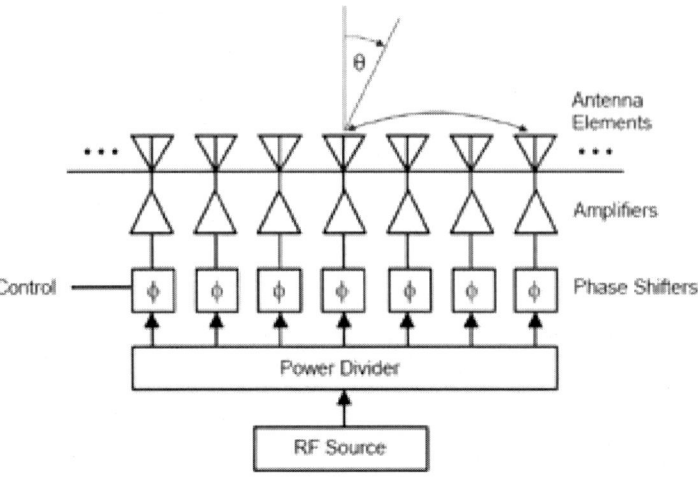

Figure 3. Adaptive antenna

The planar transverse continuous stub (CTS) was initially conceived by W.W. Milroy in 1991 at Hughes Aircraft Corporation. This reflects a modern low-cost antenna category. This technology was only recently implemented in the area of science. At microwave frequencies, it provides conventional approaches to antenna architecture. The arrangement of a CTS element array exhibits coaxial geometry from a multitude of cylindrical segments. The top and bottom ends of each of the cylindrical parts have a border. With regard to a longitudinal axis, each rim spreads transversely from the cylindrical section to form a stub portion and the actual cylindrical components are aligned end-to-end to form a coaxial cable system circling a central core material. Reactive or radiating components for microwave, millimetre-wave and quasi-optical and antennas are developed in the sequence of these stubs. Compact scale, light weight, minimized loss and strong directivity are the advantages of CTS structures. Over waveguide or patch antennas, CTS technology often has greater tunable bandwidth. Higher performance and polarization separation can be achieved utilizing this technology. In the implementation of coaxial geometries, both of the details above are beneficial. The unidirectional radiation pattern, perpendicular to the transmission line is given by the coaxial CTS structure. Another requires easier balancing of impedance, thereby ensuring greater performance and promoting alignment of the device with other systems. CTS technologies may be used to build a single-band or multiband antenna. There are three approaches of optimization to maximize the efficacy and the benefit of the coaxial CTS antenna that works on the appropriate band. These basic low-cost coaxial frameworks may be modified to accommodate wireless networking applications for microwave frequency, satellite application systems, and could be used in Friend-or-Foe systems as well. In wireless networking, the double frequency coaxial CTS antenna, working at two separate frequencies, may be modified to accommodate the base station applications. The CTS antenna has enough bandwidth. It may also be found in the upper frequency region for satellite networking and personal communication networks. The CTS antenna could be built to act as a way for the military to detect Friend-or-Foe systems. The correct separation by the coaxial transmission line between the arrays of various bands forms a single array antenna which can radiate at low and high frequencies.

5. Communication Antenna

With their exponential development, microstrip patch antennas have drawn the interest of most researchers largely because they are reasonably inexpensive to generate and because of their basic two-dimensional form, they are simpler to build. It is planned to provide a small microstrip-fed patch antenna with a patch providing a rectangular slot with the microstrip line running symmetrically to all sides with a partial ground plane. An antenna with a large slot fed by a microstrip line with a structure-like tuning stub. By joining a rectangular patch and a triangular patch, the antenna is designed. In multiple frequency bands used mostly for Wi-Fi, Bluetooth and GSM, this newly created patch helps make the antenna resonate. The rectangular patch antenna is constructed using numerous slots and shorting pins, providing characteristics of the antenna triband. From 0G to 4G, mobile cellular infrastructure has been tested, introducing the revolution in mobile telecommunications. There are many implementations of 4G technology, such as remote host control, video call data flow and connectivity of the system type. 4G has some benefits, but the issues of low efficiency, poor reach, communication interruption; heavy energy usage, poor interconnectivity and overcrowded networks will not be fixed. Present 4G infrastructure could not satisfy the new requirement because of the exponential rise in mobile devices in the networking system. To satisfy the potential need for large data speeds, the mobile connectivity infrastructure must now be updated to the next generation (5G). For 5G mobile networks, development is ongoing and is anticipated to be commercialized early in the 21st century.

With a rapid growth in consumer equipment, the demand for bandwidth is often expanded to allow vast volumes of data to pass. A range of devices such as HD TVs, tablets, mobile phones, home appliances, video monitoring systems, alarms, connected devices; cameras and robotics are projected to grow dramatically in the foreseeable future. For such devices, 5G technology guarantees the requisite data rate and capacity, such that 5G mobile connectivity is supposed to serve a vast range of real-time applications, touch internet and consistency (QoS) services (such as bandwidth, jitter, packet error, latency and packet loss) and quality of experience (such as network operators and users). The key advantages of the 5G infrastructure are high-speed data processing, zero latency and universal access. 5G technologies can link both electronic and digital appliances/services, such as temperature maintenance, scanners, air conditioners, refrigerators, door locks for LEDs, microwave ovens, etc. and will enable such appliances to be remotely operated. 5G technology is known as a true, unregulated digital environment that will create a global wireless platform and interactive ad-hoc wireless network. It also includes high quality and bidirectional broad giga-bit bandwidths to render the advance billing system more appealing and usable with the exclusive provision of several data transmission routes is shown in Figure 4.

Figure 4. Communication antenna.

In a modern period of automated mobile communications networks, 5G technology was launched, which brings the Internet of Things (IOTs) into the Smartphone network (D2Ds) with broader capacity, lower power usage, higher data speeds and greater coverage, which can address multiple problems such as latency, stability, efficiency, affordability and cost of equipment. One of the key issues of 5G communications is bandwidth sharing. The 5th century of the millimetre wave band is being researched for mobile communications. For 5G mobile communications, some of the planned bands suggested are: 27.5-29.5 GHz, 33.4-36 GHz, 37-40.5GHz, 42-45 GHz, 47-50.2GHz, 50.4-52.6 GHz and 59.3-71 GHz. However the 28GHz and 38GHz frequencies are beneficial for wireless communications, but working at higher frequency bands would add complexity to the mobile communication device antenna architecture. Due to the higher data rate and directional radiation for wireless devices, microstrip patch antennas are of considerable importance. For the first time, Heinrich Hertz showed a patch antenna in 1886, and in 1901, Guglielmo Marconi provided its practical use. The biggest problem for antenna manufacturers is the bandwidth and scale of the antenna for realistic applications. Microstrip patch antennas have various benefits such as low cost, compact size, high performance, fast processing and manufacturing, and when installed on a rigid surface, they are mechanically durable. By utilizing multiple feeding strategies and shapes of radiation pads, dual and triple band operating frequencies can be effectively accomplished with limited efforts along with separate polarizations of E-fields. Global Positioning System (GPS), cellular local area networks, handheld networking devices, and microwave sensors are commonly used for microstrip antennas.

Microstrip antennas still have several drawbacks, which involve restricted bandwidth and poor gain, considering these desirable advantages. Analysis demonstrates that bandwidth can be increased using numerous strategies such as slotted patch, dense substrates, low-effective permittivity substrates, several resonances combined, and impedance matching optimization. To maximize the gain constraints, array design is applied. Some researchers have recently put forward different patch antenna designs, especially for the fifth generation of mobile communications. The 5G PIFA antenna of 28GHz and 38GHz operating frequencies comprises of a short patch with a U-shape slot etched in the radiating patch. At 28GHz, the maximum gain achieved is 3.75 dB and at 38GHz, 5.06 dB. To separate the field and the patch, a FR-406 substrate is used. The efficiency and architecture sophistication for 5G mobile phone networks of the microstrip patch antenna. The antenna is a transducer for the propagation or receiving of electromagnetic waves (EM). Compared to traditional antennas, the microstrip antenna has many benefits. It consists of a dielectric substratum radiating patch on one side and has a land plane on the other side. Due to their light weight and low expense, microstrip antennas are used for many commercial purposes. The latest specifications of portable wireless devices are driving the need for reconfigurable pattern antennas. Compared to traditional antennas providing one feature in a single antenna, the reconfigurable microstrip antenna offers multiple applications and offers further flexibility. They will offer adaptive variation of mobile communications in operating frequency, radiation pattern and polarization.

There are different forms of feeding strategies available for microstrip antenna feeding. Each of them has strengths or demerits of their own. In order to select which type of feeding is sufficient for the constructed antenna, a variety of variables are used. The primary concern is the efficient transmission of power from the feed line to the radiating portion of the antenna that is correctly aligned between the feed and the antenna. For impedance matching, different methods are used, such as impedance transformer, stubs. The feed framework should be able to efficiently produce these matching structures with radiating components. The key variables that depend on the feeding techniques that influence the antenna features are also spurious feed radiation and surface wave losses. Surface waves reduce antenna performance and spurious feed radiation results in undesired radiation, contributing to the level of the side lobe and raising the level of cross polarization as well. Another key aspect is that the feed network should be well-suited to build a collection; feeding strategies can be split into two groups, one touching feeds and another non-contacting feeds or electromagnetic coupled feed. The feed line is directly linked to the radiating portion when the feed is contacted.

These methods include the use of cross slots and sorting sticks, the use of circular and triangular patches with correct slits and antenna arrays to maximize the thickness of the patch. Various feeding strategies are often widely researched to address these limits. In recent wireless communication technologies, dual frequency function of the antenna has become a requirement for many applications. They have become appealing for many applications owing to the low profile, light weight and low development costs of microstrip patch antennas. Wide bandwidth and compact, low-profile antennas are needed for new trends in communication systems. On a thin dielectric substrate, microstrip patch antennas naturally have the downside of short bandwidth impedance. In aircraft, satellite and missile applications, where heights, weight, expense, efficiency, ease of installation, and aerodynamic profile are strict restrictions, these low profile antennas are also useful. Low profile character, conformability to, is some of the key advantages of this kind of antenna. A big disadvantage of these antennas though is the narrow bandwidth. Different attempts by researchers have been made to expand its bandwidth. Ultra-wideband (UWB) is an emerging radio technology that has recently gained a lot of coverage. Ultra wideband (UWB) communication systems may be generally categorized as any communication device with an immediate bandwidth several times greater than the minimum needed for precise information to be delivered. The fundamental drawback of small impedance bandwidth is to include all current wireless networking networks such as AMPC800, GSM900, GSM1800, PCS1900, WCDMA/UMTS (3G), 2.45/5.2/5.8-GHz-ISM, UNII, DECT, WLAN, European Hiper LAN I, II, microstrip patch antennas on a thin dielectric substratum. So we use a co-planar feed to maximize the antenna bandwidth. The patch is positioned next to a tiny rectangular floor co-planar to it unlike the normal way of positioning the radiating patch of the microstrip antenna on top of a ground plane. Integrated microwave circuits (MIC) and monolithic integrated microwave circuits (MMIC) can be conveniently integrated with them. The substitution of coaxial feeding or line feeding with coplanar feeding is a basic but effective technique.

6. Ultra Wideband Antenna

In the current situation, wideband circularly polarised microstrip patch antennas in a broad variety of wireless networking applications such as radar and bio-medical telemetry devices draw a great deal of interest. Because of their many fascinating characteristics, such as small design, low profile, light weight, ease of production and connectivity with their microwave parts, these apps have been feasible. The biggest downside of microstrip patch antennas is their very short bandwidth, so the analysis was carried out to raise the gain and impedance of the patch antennas by preserving the patch antenna scale as rapidly as possible. A microstrip patch antenna is a community of several active antennas to generate a directive radiation pattern and gain, coupled to a standard source or load. The circularly polarised microstrip patch antenna in many contact systems is used as powerful radiators. This patch antenna has high gain, increased performance and a good pattern of radiation. The industrial use of the frequency spectrum for the ultra large band is 3.1 GHz to 10.6 GHz.

Generally, a wideband antenna is commonly used to design the FR-4 substrate. A thicker substrate is used to improve the bandwidth of a patch antenna and a microstrip line feeding technique is therefore necessary. The ultra-wide band technology is chosen for highly reliable, low cost and low complexity purposes. It is used in numerous uses, such as radar, aircraft and military contact. Different techniques are used to increase the quality of the gain. Two orthogonal dipoles of separate lengths were initially designed for the antenna. A rectangular patch engraved on a flame retardant (FR-4) substrate with a 50 ohm feed line consists of this planar ultra broad band antenna. With one slot in the ground plane, the rectangular patch has one round cut at each corner. This is done by inserting slots in the patch, feed and ground plane of varying shapes or utilizing faulty ground structure (DGS). The effects of the calculation agree with a small change in the lower and upper and edge frequencies. In wireless networking and radar device systems, microstrip antenna (MPA) configurations have various benefits. This is due to the low weight, low profile, low expense, surface-compatible, mechanically stable and versatile polarization, frequency, pattern & impedance. Through etching metal on one side of the dielectric substrate, a patch antenna is formed where there is a continuous metal layer of the substrate creating a ground plane on the opposite side. MPAs are generally low performance narrowband, low power and high Q antennas, but multiple strategies for optimizing bandwidth are needed in making the size as small as possible to be perfectly matched as a low-profile antenna. Because of the reality that many experiments and study are being carried out all around the world. MPA bandwidth is basically narrow, but wireless networking networks today need higher operational bandwidth.

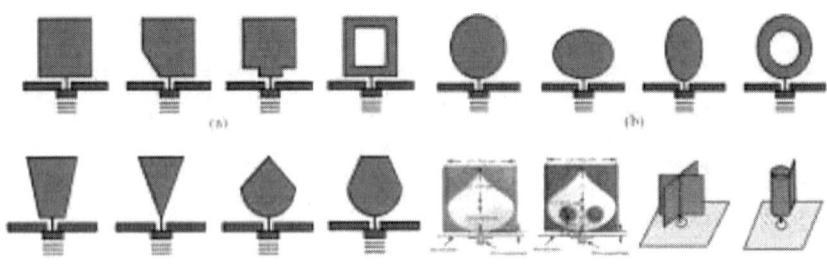

Figure 5. UWB antennas.

The need for smaller and low-profile antennas has taken the Microstrip Antennas (MSA) to the fore, with the increased needs for personal and mobile communications is shown in Figure 5. Since MSA is known to have properties such as low size, weight, expense, power, installation ease, these antennas have very good output. Because of the radiating patch on one side of the dielectric substrate and the field on the other side, MSA is also known as the patch antenna. For use in cell phones, personal Bluetooth networks and wireless local networks, the physical characteristics of MSA are compatible. Although these antennas have many advantages, they also suffer from some disadvantages such as limited performance, low strength, and bandwidth of very narrow frequency. Rectangular MSA is the simplest and most widely used MSA. The MSA struggles from the biggest downside of insufficient bandwidth. Researchers have carried out a variety of approaches to address the shortcomings of MSA. By increasing the permissibility of the substrate or by increasing the height of the patch, the shortcomings of MSA may be resolved. However because raising patch height results in changing the low-profile function of the patch, the results of each of these approaches are insufficient, whereas the application of the second is subject to material supply and suitability.

The usage of parasitic elements lying on the same sheet or another (stacked) layer is another strategy. The aperture coupled excitations were often used to obtain the broader bandwidth, but the development method is complicated by this approach. The different slot shapes are required to meet the bandwidth upgrade criterion. Microstrip antennas have the benefit of low expense, thin profile, light weight, ease of fabrication, conformable to mounting surface and being incorporated in active products. Probe fed microstrip antennas also have outstanding separation between the feed network and the elements of radiation and produce very strong front-to-back ratios. Because of these benefits, microstrip antennas are used in a number of areas, such as space exploration, aero planes, rockets, surveillance, cell phones, GPS, remote sensing, and satellite transmission. Narrow bandwidth is the most significant downside of microstrip antennas. With classical microstrip antennas, a limit of 8 percent bandwidth is usable. The market for small, low profile and broadband antennas has risen dramatically with the wide proliferation of wireless networking technologies in recent years. Because of its low profile, light weight and low expense, the microstrip patch antenna was a way to satisfy the market. Microstrip patch antennas running in the millimeter-wave spectrum are becoming increasingly popular in Bluetooth, Wi-Fi, WLAN and many other wireless networking implementations to reach emerging hardware niches. Conventional microstrip patch antennas, however, suffer from less radiation quality, limited bandwidth, poor directivity and gain due to lack of surface wave and layout of large and thicker substratum dimension. This raises a technological problem for the manufacturer of the microstrip patch antenna to conform to broadband techniques.

Table 1. Advantages and disadvantages of microstrip patch antenna

Sl.No	Advantages	Disadavantages
1.	Low Weight	Low Efficiency
2.	Low Profile	Low Gain
3.	Linear and circular polarization.	Surface Waves Excitation
4.	Dual and Triple Operating Frequency	Polarization purity is difficult to achieved
5.	Planar feeding Higher bandwidth	Fabrication Alignment is important for multilayer input match

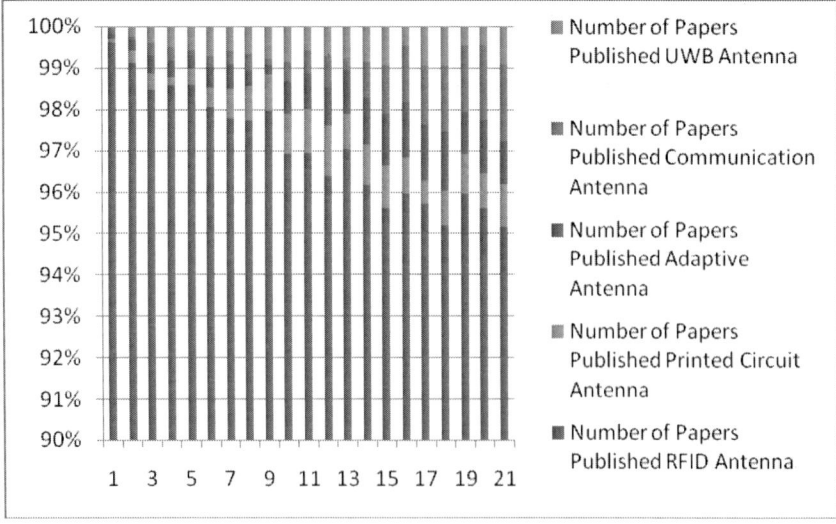

Figure 6. Comparison chart for various field of antenna.

Table 2. Number of papers published in microstrip patch antenna

Year	Number of Papers Published				
	RFID Antenna	Printed Circuit Antenna	Adaptive Antenna	Communication Antenna	UWB Antenna
2000	1	2	2	3	1
2001	2	6	5	2	5
2002	5	8	9	6	8
2003	8	4	8	7	10
2004	6	8	7	2	12
2005	5	10	10	5	15
2006	8	15	12	7	12
2007	6	17	8	8	14
2008	5	18	5	3	16
2009	5	20	16	10	18
2010	8	22	18	12	12
2011	9	26	19	16	15
2012	5	18	22	6	16
2013	7	21	24	18	18
2014	8	22	26	25	20
2015	10	18	28	29	10
2016	15	12	29	30	20
2017	20	18	30	35	20
2018	22	20	22	34	10
2019	24	18	28	38	10
2020	26	22	22	40	20

Several approaches have been suggested to increase the quality of radiation in state-of-the-art antenna testing. The newly recorded methods to improve the radiation performance and the directivity of patch antennas are the use of thick substrate, low dielectric substrate, multi-resonator stack arrangements, impedance balancing, slot antenna geometry and cutting a resonant slot within the patch. Like co-axial feed, aperture coupling, proximity coupling, several methods of microstrip antenna feeding. Due to its clear processing, fast attachment to dielectric substrate and impedance matching property, microstrip line feeding is an effective feeding scheme among the techniques. The dielectric substrate's thickness, however, deteriorates the performance of the antenna bandwidth by growing surface wave and spurious feed radiation along with line feeding. Consequently, feed radiation results are driven by an undesired cross polarized radiation. The wireless communication industry is therefore focused on improving the architecture of the microstrip patch antenna to achieve full radiation quality. We also planned and studied the radiation of a microstrip antenna with a rectangular patch in this article. The effects of the dielectric constant and substrate height on antenna radiation efficiency were studied using numerical analysis and MATLAB simulation methods. Finally, to reserve adequate radiation efficiency, a collection of optimum architecture parameters was suggested. The microstrip patch antenna has been various advantages and disadvantages are shown in Table 1. The analysis of various applications for Microstrip patch antenna are shown in Table 2 and Figure 6.

7. Conclusion

An antenna is a transducer that transforms alternating current or vice versa into radio frequency fields. In the modern world, wireless communication has evolved widely and rapidly, particularly during the last decade. The creation of personal connectivity systems would seek to include images, DMB (Digital Multimedia Broadcasting), video telephony, voice and data communication at any point in the immediate future, everywhere in the world utilizing Wireless Local Area Networks (WLANs). The specifications for miniaturized multiband antennas with acceptable frequency bands ideal for Wi-Fi (IEEE 802.11 standard) and WiMAX (IEEE 802.16e-2005 standard) implementations have been ignited by rapid developments in different WLAN protocols. Wi-Fi is used in the 2.4 GHz (2.4 GHz-2.5 GHz) and 5 GHz (5.15-5.35 GHz, 5.47-5.725 GHz and 5.725-5.875 GHz) bands, respectively. The operating bands for WiMAX are 2.3 GHz (2.3-2.4 GHz), 2.5 GHz (2.5-2.7 GHz) and 3.5 GHz (3.4-3.6 GHz).

Chapter 3

Microstrip Patch Antennas Using Various Structures for Various Applications

Abstract

Over the past decade, there has been a massive increase in the number of wireless and mobile users all over the globe. Wireless technology is being used for a wide range of applications, including Internet and web surfing, video, and other text and multimedia applications. The tendency in radio frequency is to migrate from narrow band to broad band. Most embedded systems nowadays are portable devices, and these portable gadgets rely on mobility. Embedded communication plays a crucial role in communicating with these portable devices. An effective antenna device capable of transmitting bit streams with minimal return loss, high data rate, high directivity, and bandwidth is a crucial need of such a system. An efficient microstrip antenna design is required to achieve a broad bandwidth. Micro strip patch antennas are ideal for high-performance aeroplanes, spacecraft, satellites, missiles, and embedded applications because to their small weight, low profile, cheap manufacturing cost, dependability, and simplicity of fabrication and integration with wireless technology equipment. They do, however, have a poor radiation efficiency, a low power output, a high Q, and a very restricted frequency spectrum. The form of the radiating patch is used to name microstrip patch antennas. Radiating patches come in a variety of forms, including square, rectangular, circular, elliptical, triangular, circular ring, and ring sector. Microstrip patch antennas in square, rectangular, and circular shapes are simple to construct and analyse. They are more prevalent as a result of these characteristics. Circular microstrip patch antennas are simpler than rectangular microstrip patch antennas because they only have one degree of freedom to adjust (radius), while rectangular microstrip patch antennas have two (length and breadth). As a result, circular microstrip patch antennas are easier to build and regulate their emission. Furthermore, at

In: Recent Trends in Microstrip Antennas for Wireless Applications
Editors: S. Kannadhasan and R. Nagarajan
ISBN: 978-1-68507-744-0
© 2022 Nova Science Publishers, Inc.

the same design frequency, the circular patch antenna has a physical dimension that is 16% less than the rectangular microstrip antenna.

1. Introduction

A microstrip patch antenna features a radiating patch on one side of the dielectric substrate and a ground plane on the other, with the radiation coming from the fringing fields between the patch's perimeter and the ground plane. Thicker substrates with lower dielectric constants will result in improved antenna efficiency, bandwidth, and performance. They do, however, result in bigger antennas. Thin substrates with high dielectric constants, on the other hand, are ideally suited for microwave applications due to their tightly confined fields, which result in less unwanted radiation and coupling. It has a lot of advantages because of its smaller size, but the losses are larger, making it less efficient and resulting in narrow bandwidths. A microstrip patch antenna may be fed in a variety of ways. Microstrip line, coaxial probe, and aperture coupling are the most often utilised. The approaches may be divided into two categories: contacting and non-contacting. RF power is directly delivered to the radiating patch through a microsrip line in the contacting approach. To transmit power between the microstrip line and the radiating patch in the non-contacting situation, electromagnetic field coupling is utilised. A microstrip line feed is employed in the proposed circular patch antenna design.

Microstrip line feed is a contacting technique that uses an extremely thin conducting strip relative to the patch width (width should be less than that of patch thickness). Because the antenna's impedance is built at 50 Ohm, it's straightforward to construct, match, and design. The location of the feed line in the patch should be modified to fit the impedance. Different slots may be carved within the patch shape to improve the antenna's performance. The antenna characteristics of a modified circular microstrip patch antenna with circular slot were studied using the HFSSTM simulator software.

Nowadays, communication plays an essential role in global society, and communication systems are fast transitioning from wired to wireless. Telecommunications strives to achieve the highest levels of performance, reliability, and efficiency at the lowest feasible cost. Antennas serve as a fundamental element in this sector, enabling electromagnetic waves to be transmitted in free space. There are various kinds of antennas, each with its own cut, geometrical form, and transmission capability. Microstrip patch antennas have a low profile, meaning they are thin and simple to manufacture, which gives them a significant advantage over regular antennas. Planar antennas used in wireless networking and other microwave applications are known as patch antennas. Microstrip is a planar technology for creating lines that carry communications and antennas that couple such lines with radiated waves. A patch is usually broader than a strip, and its form and size are essential antenna characteristics. When using microstrip patch antennas, microstrip antennas are very useful. High loss and surface waves in the substrate layer are issues that will arise, since losses will always occur in the radiation when the antenna transmits the signals.

Due to features such as small weight, low volume, cheap cost, compatibility with integrated circuits, and ease of installation on a hard surface, microstrip patch antennas are arguably the most extensively used form of antenna today. They may also be simply configured to work in dual-band, multi-band, dual, or circular polarisation. They play a key role in a variety of commercial applications. However, since microstrip patch antennas have a limited bandwidth by design, and bandwidth augmentation is often required for practical applications, a variety of ways have been used to increase bandwidth. We designed a circular microstrip antenna with a dielectric constant of 4.4 and a copper thickness of 50 microns, utilising FR4 epoxy for microstrip antennas at various frequencies (2, 4, 6, 8and 10GHz). Because of its inexpensive cost and easy availability, FR4 was selected for this investigation. It may therefore be utilised for prototyping microstrip antenna arrays. DGS is achieved by inserting a defective form on a ground plane, which disrupts the shielded current distribution based on the defect's shape and size. The input impedance and current flow of the antenna will be affected by the disturbance at the shielded current distribution. It may also regulate the excitation and propagation of electromagnetic waves via the substrate layer. Because surface waves may extract all available power for radiation to space waves, losses due to surface wave stimulation will result in a drop in antenna efficiency, gain, and bandwidth. As a result, the bandwidth of a microstrip antenna without DGS is limited, and the return loss is considerable. Microstrip antennas with DGS, on the other hand, will have a greater working bandwidth and lower return loss. As a result, the DGS may be incorporated into the ground plane to boost bandwidth. Microstrip antennas may be designed using a variety of feeding approaches. At 50 MHz, we employed the co-axial feeding approach, which is often used for appropriate impedance matching.

Low-profile, light-weight, low-cost, easy-to-fabricate, and compact radiators are required in today's wireless communication era. The most ideal antenna for this application is a slot or patch antenna. However, these antennas (in their traditional form) have significant drawbacks, including poor gain and restricted bandwidth. To eliminate these flaws, a lot of study has been done. The intrinsic aspect of slot radiators is that they may produce an omnidirectional pattern in one of their major planes, which is not achievable with traditional patch antennas. As a result, slot antennas are preferred over microstrip radiators. Recently, a lot of attention has been paid to multiband antenna properties. It's because of two major factors: I A single antenna may cover various frequency bands; (ii) improved signal-to-noise ratio due to band rejection. To get dual-band characteristics, multiple strategies were employed at first, such as using different form slots, such as a cross slot, a circular slot, a square slot, and an offset circular slot]. In 2011, a dual-frequency microstrip antenna was suggested. An offset microstrip-fed line and a strip adjacent to the radiating edges in the circular slot patch were used to achieve impedance bandwidths of 26.2 percent and 22.2 percent at lower and higher operating bands, respectively. It is bigger than the suggested antenna construction (108 108 1.6mm3). For dual-band applications, a helix-shaped slot radiator was suggested. This radiator can be used in the GPS L1 and L2 frequency bands, although it has a big antenna. A cavity-backed slot antenna for dual-band applications was suggested in 2016. The notion of a spurline complicates antenna design. A split ring resonator loaded square slot antenna for dual-frequency applications was suggested shortly after. In terms of bandwidth and gain, this antenna performs well, however it suffers from a huge antenna size. A simple dual-band slot antenna design was recently presented. It has two rectangular slots that are offset from one another. WLAN and WiMAX applications are also compatible with this antenna. However, it is bigger than the intended antenna construction (60 60 1.6mm3).

2. MPA Various Structures

A modified circular shaped slot antenna for dual-frequency applications is shown in this article. The proposed slot antenna is fed by an asymmetrical microstrip line. This antenna configuration supports two frequency bands: 2.88–3.92 GHz and 5.26–6.28 GHz. With typical gains of 3.0 dBi and 6.0 dBi in the lower and higher frequency regions, it delivers steady far-field characteristics. Microstrip patch antennas provide a number of practical benefits, including a low profile and minimal weight. They can be polarised in both linear and circular directions, however they have a low gain and a restricted bandwidth (usually 1–5%), which is a key limiting factor for their widespread use. This study reviews a microstrip antenna using various strategies such as slotted antennas, stacking, alternative feed mechanisms, slits, and truncation to solve these drawbacks. Microstrip antennas may be employed in a wide range of applications by overcoming these limits, with RFID scanners being one of the most common. RFID (Radio Frequency Identification and Tracking) is a technology that allows for wireless identification and tracking. RFID technology has quickly evolved and been deployed to various service sectors, distribution logistics, and manufacturing organisations in recent years. The reader emits signals using reader antennae in an ultra-high-frequency (UHF) RFID system. An antenna and an application-specific integrated circuit make up an RFID tag at the client (ASIC). The reader activates the tag and queries it for information about its content. The reader's query signal must be strong enough to activate the tag ASIC, which will execute data processing and send back a modulated string across the specified reading distance. Circularly polarised (CP) reader antennas have been utilised in UHF RFID systems to ensure the reliability of communications between readers and tags since RFID tags are always freely oriented in real use while tag antennas are generally linearly polarised. The loss produced by multipath effects between the reader and the tag antenna may be reduced using circularly polarised microstrip antennas. In wireless communication systems, circularly polarised (CP) radiation is useful since it operates regardless of the orientation of the transmitter and reception antennas.

The UHF band has a total frequency range of 850 MHz to 960 MHz, which is used for international RFID systems. The RFID system uses frequencies ranging from 902 to 928 MHz in the United States, 865 to 867 MHz in Europe, and 840 to 955 MHz in Asia-Pacific. China (920.5 MHz to 924.5 MHz), Japan (952 MHz to 955 MHz), India (865 MHz to 867 MHz), Hong Kong (865 MHz to 868 MHz and 920 MHz to 925 MHz), Taiwan (920 MHz to 928 MHz), Korea (908.5 MHz to 910 MHz and 910 MHz to 914 MHz), Singapore (866 MHz to 869 MHz), and Australia (866 MHz to 869 MHz) (920 MHz to 926 MHz). For global RFID applications, a circularly polarised reader antenna must cover the whole UHF band (850 MHz to 960 MHz).

This document contains a variety of strategies and geometries that have been created and implemented by many authors in order to enhance bandwidth and generate high-quality circularly polarised radiations. One of the ways is the single feed, stacked, slotted circularly polarised microstrip antenna, which uses two circular slots to create circular polarisation. For UHF RFID applications, a wideband circularly polarised antenna with three stacked patches and a hybrid coupler was devised. Asymmetric-slotted microstrip square patches have recently been developed to reduce the size of circularly polarised microstrip antennas. However, these antennas have a restricted circularly polarised radiation bandwidth, low gain, and are intended for hand-held or portable RFID readers, which are disadvantages. When compared to a narrowband tiny circularly polarised antenna (low gain) used in a handheld or portable reader, a wideband circularly polarised RFID antenna with high gain for a stationary reader has a larger reading range. The circularly polarised antenna's performance is additionally improved by a stacking ring above the radiating patch is shown in figure 1. The primary patch is fed by four probes, which are progressively linked to the suspended microstrip feed line, to create a wideband circularly polarised stacking antenna.

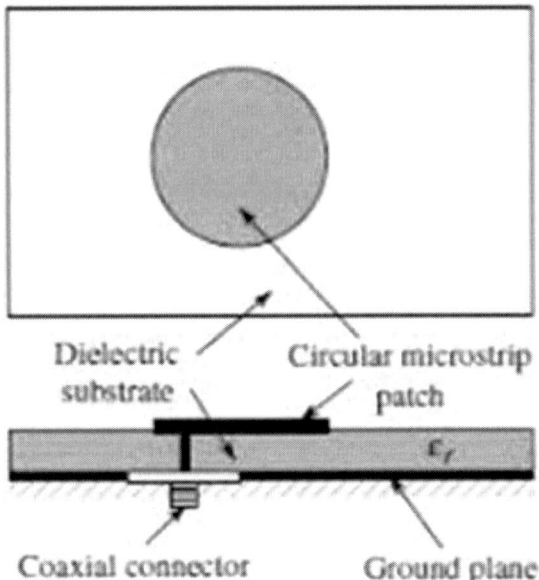

Figure 1. Circular patch antenna.

When compared to a simple juxtaposition of antennae and solar cells, the notion of integrating photovoltaic solar cells with microwave antennas provides a broad variety of benefits in terms of surface coverage, volume, mass, cost, and electric performance. Communication systems that combine photovoltaic technology for low-cost and stand-alone applications have recently attracted a lot of attention. When solar power generating systems are integrated with communications systems, compact and dependable autonomous communication systems may be created for a variety of applications.

Commercial solar cells are attached or positioned close to the radiating patch or in the ground plane of slot antennas in the majority of documented efforts at solar cell integration with printed antennas. Other arrangements, such as putting the solar cells behind the reflect array antennas, have been considered. Following the creation of an amorphous-Si cell on a flexible thin film polymer substrate, better photovoltaic performance was achieved at a cheaper cost. The combination of amorphous silicon solar cells with microstrip slot antennas allowed for a greater degree of integration. The slot antennas were chosen to reduce the influence of solar cells on the antenna's RF performance. However, this slot antenna has limitations such as a restricted bandwidth and poor circular polarisation performance. To create the correct shapes of slots, complex laser cutting of the solar cells is necessary. Because complicated slot designs in solar cells are difficult to etch, designing dual-frequency and multi-frequency solar slot antennas is tough.

Furthermore, the enormous size of the slot structures will reduce the efficiency of the solar cells. For WLAN functioning, printed slot antennas have received a lot of attention. To excite the dual resonant modes, the key to providing a flexible design is to make the slot in the solar cell as narrow as feasible with a simple geometric form. ML has been extensively explored as a complementary technique to CEM in designing and optimising different kinds of antennas15-18 for multiple benefits, as will be described further in this study, due to their intrinsic nonlinearities. ML is a broad subfield of artificial intelligence that focuses on extracting meaningful information from data, which explains why it's been linked to statistics and data science so often. Indeed, machine learning's data-driven approach has enabled us to design systems like never before, bringing us closer to creating completely autonomous systems that can equal, rival, and occasionally surpass human talents and intuition. However, the quality, amount, and availability of data, which may be difficult to collect in certain situations, is critical to the success of ML techniques. This data must be gathered, if not already accessible, from the standpoint of antenna design, since no standardised dataset for antennas, such as those used for computer vision, is currently available. This may be accomplished by utilising CEM simulation software to simulate the desired antenna across a broad range of values. A dataset may be produced and separated into train, cross-validation, and test sets based on the acquired findings for the aim of training an ML model and evaluating if it succeeds in generalising on additional inputs. It's up to the designer's foresight and knowledge at this stage to figure out how to diagnose the model and enhance performance.

An antenna array is a set of two or more antenna components that may be arranged in a certain pattern. An antenna element is positioned along one axis in a linear antenna array. The geometry (linear, circular, spherical, etc.) and several other factors, such as inter-element spacing, excitation amplitude, and excitation phase of the individual element, all affect the beam produced by the antenna array. In most cases, wireless communication necessitates the use of a more directed antenna with a high gain. Antenna arrays have a higher gain, are more directional, and have more spatial variety. In the recent past, antenna array synthesis has gained popularity, with different performance targets being examined. Various optimization techniques, such as simulated annealing, ant-colony optimization, GA, and PSO, have been applied in past research investigations for array synthesis. By combining the PSO with the GA, the linear antenna optimization produces a pattern with the lowest SLL and HPBW. To reduce the SLL and position nulls in the desired direction, the composite differential evaluation (CoDE) technique was used to improve inter-element spacing between two consecutive elements.

In the optimization process, a multi-objective optimization strategy was utilised to optimise the directivity and reduce the SLL of an antenna array. The authors presented the memetic generalised differential evaluation (MGDE3) algorithm, a novel memetic multi-objective evolutionary algorithm that is an extension of the generalised differential evaluation (GDE3) method. The authors presented a method for realising the non-uniform linear array's features and performance. Multi-objective functions were used in the described approach to optimise inter-element spacing, excitation currents, and excitation phases, as well as reduce SLL and HPBW. By adjusting spacing and excitation amplitude, real-coded genetic (RCG) was used to impose nulls in the desired direction on a time modulating linear antenna array.

3. Applications of Microstrip Patch Antenna

Wireless applications may be found in the construction and design of a u-slot microstrip antenna. Wireless communication systems are becoming widely used since they are less costly and provide a flexible and alternative communication method. The suggested antenna is based on a space-saving design. In this case, optimizations aren't better than doing nothing. It's also necessary to look for any programmes that may be improved. This optimization was reduced to a space decrease and occurred during the software compilation. The suggested microstrip patch antenna comes in a variety of forms, such as U, E, I, and T, among others. Different shapes and radiating patches were created using the DGS algorithm in this work. The antenna gain is 2.58 dB, and the dimensions are 65mm*62mm*105mm. After that, the impedance bandwidths are compared to those of another patch antenna. It features excellent radiation, a decent Bluetooth range, and a wireless communication application. Microstrip antennas provide the following advantages: cheap cost, light weight, and low volume. The notion of equivalent circuit is used to study a microstrip u-slot patch antenna. The influence of base length, u-slot width, and arm length is explored using the parametric suggested approach. Satellite communication, medicinal services, mobile communication, aviation, and radar are just a few of the uses. In the suggested technique, the theoretical simulation results are provided. The Bluetooth application and wireless communication are discussed in this work. Bluetooth technology is fast expanding in the fields of mobile phone charging, profile, interference, and identification. The suggested technique may include industrial science and medical methods. The U-shape microstrip antenna is intended for use at 2.4GHz. It's popular because of its low weight and low profile, however it has downsides such as poor gain and restricted bandwidth. The ground regulator is used in a microstrip patch antenna with a double band u-slot configuration. The gain ranges from 1.786 to 4.80 dB, with directivity ranging from 6.50 to 6.6 dBi. Because of its low profile and low mismatch loss of 0 to 10, communication technology has been a key feature of the u-slot structure. The goal of this suggested two-stage optimization advance is to quickly organise a worldwide point of optimum region, and then apply optimization to the global point of optimal accuracy. The benefits increase the suggested method's efficiency, simulation, and based equation.

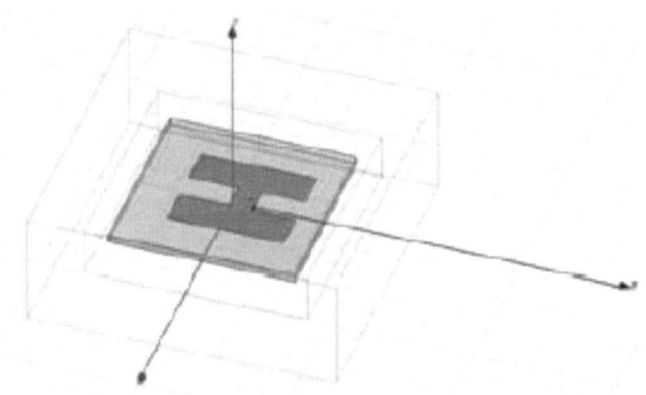

Figure 2. H shaped structure.

This article simulates and designs a single u-slot micro-strip patch antenna with a 5.6GHz operating frequency. The antenna's patch size was reduced to a rectangular shape, which enhanced the antenna's return loss and gain. Omni-direction is a technology that transmits and receives data in a horizontal plane. The proposed solution involves the use of 5G technology and the creation of a u-slot rectangular patch array. The current technique may be implemented as a 5G application with many inputs and outputs. Antenna with a frequency of 27 GHz and a duration of 2 seconds. This article depicts a Micro-strip patch antenna that operates in the L band for satellite communication at a frequency of 2.5 GHz. One of the most common antenna types is the microstrip patch antenna, which has two sides and has been frequently used. The ground route is on one side and the radiating path is on the other. Low cast, light weight, tiny size, and low profile are all benefits. The investigation was carried out by providing four semi-circular U-shaped slot structures that were based on a theoretical specification. With the comparable LC lumped parameters in charge of providing the notched frequency, the formulation is accepted. By analysing the C and L proportional formula, the LC equivalent notched frequency was developed, and it was accepted using simulated and estimated results.

Bluetooth technology establishes short-range wireless connections between electronic devices such as computers, mobile phones, and other devices, allowing audio, data, and video to be exchanged. Due to the fast advancement of communication standards, there is a high need for antennas that take up little space, have a small profile and size, are inexpensive to manufacture, and are simple to integrate into a feeding network. Because they are low weight, small, simple to integrate, and cost efficient, microstrip patch antennas are commonly employed. Patch antennas, on the other hand, have a limited bandwidth owing to surface wave losses and a big patch size for improved performance. Various techniques such as frequency selective surface, stacked configuration, thicker profile for folded shorted patch antennas, thicker substrate, slot antennas such as U-slot patch antennas with shorted patch, double U-slot patch antenna, L-slot patch antenna[8], annular slot antenna, double C patch antenna, E-shaped patch antenna, and feeding techniques such as L-probe feed, circular coaxial probe feed, proximity coupled feed, and feeding techniques such as L-probe feed, circular It's important to consider the size of the feeding patch and the thickness of the dielectric. Short circuited elements, high dielectric constant materials, slots, and resistive loading have all been presented as ways to minimise the size of the patch. However, using a slot antenna has the disadvantage of having a restricted bandwidth and poor circular polarisation performance, as well as requiring sophisticated laser cutting of solar cells to produce the correct shape during construction.

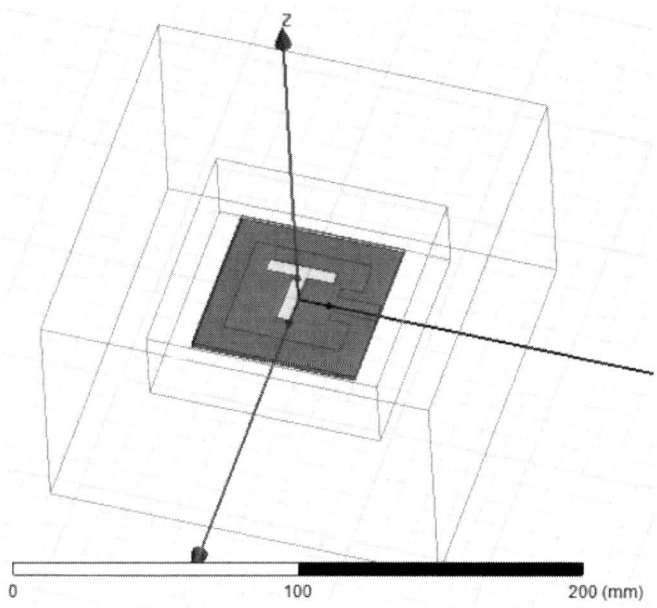

Figure 3. T shaped structure.

Antennas such as monopoles, printed monopoles, and dipoles boost bandwidth to a higher degree. Monopole antennas, on the other hand, are huge and complex to manufacture and integrate. Printed monopole antennas provide a number of benefits, including a low profile, compact size, and ease of integration, but they have the drawback of a narrow impedance bandwidth and a poor omnidirectional emission pattern. The input impedance of dipole antennas is quite high. As a result, at the feed point, an impedance matching transformer or balun coil is needed, which increases the antenna's size.

Bluetooth technology allows voice, data, and video to be shared between electronic devices such as computers, mobile phones, and other devices across short distances. Because communication standards are rapidly evolving, there is a significant need for antennas that take up little area, have a compact profile and size, are affordable to produce, and are easy to integrate into a feeding network. Microstrip patch antennas are widely used because they are light, compact, easy to integrate, and cost effective. Patch antennas, on the other hand, are restricted in bandwidth due to surface wave losses and have a large patch area for better performance. Various techniques such as frequency selective surface, stacked configuration, thicker profile for folded shorted patch antennas, thicker substrate, slot antennas such as U-slot patch antennas with shorted patch, double U-slot patch antenna, L-slot patch antenna[8], annular slot antenna, double C patch antenna, E-shaped patch antenna, and feeding techniques such as L-probe feed, circular coaxial probe feed, proximity coupled feed It's crucial to think about the size of the feeding patch and the dielectric thickness. Short circuited components, materials with a high dielectric constant, slots, and resistive loading have all been proposed as solutions to reduce the patch's size. However, utilising a slot antenna has the drawbacks of limited bandwidth and poor circular polarisation performance, as well as needing complex laser cutting of solar cells to get the necessary shape during assembly.

Monopoles, printed monopoles, and dipoles are examples of antennas that increase bandwidth. Monopole antennas, on the other hand, are enormous and difficult to build and integrate. Printed monopole antennas offer some advantages, such as a low profile, small size, and simplicity of integration, however they have a restricted impedance bandwidth and a poor omnidirectional emission pattern. Dipole antennas have a very high input impedance. As a consequence, an impedance matching transformer or balun coil is required at the feed point, increasing the antenna's size.

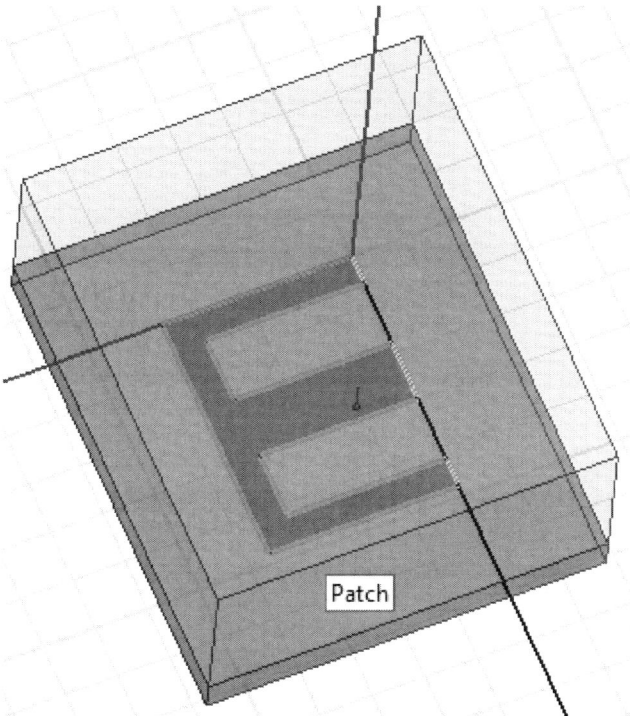

Figure 4. E shaped structure.

A transducer that transforms radio frequency fields into alternating current or vice versa is known as an antenna. For sending or receiving radio communications, there are both transmitting and receiving antennas. The antenna is critical to the functioning of any radio equipment. Antennas are used in radio broadcasting, broadcast television, two-way radio, communications receivers, radar, mobile phones, satellite communications, and other technologies. An antenna is a collection of conductors (elements) that are electrically linked to the receiver or transmitter. The oscillating current provided to the antenna by the transmitter produces an oscillating electric and magnetic field around the antenna components during transmission. As a moving transverse electromagnetic field wave, these time-varying fields emit energy out from the antenna into space. The oscillating electric and magnetic fields of an incoming radio wave, on the other hand, exert stress on the electrons in the antenna components, forcing them to travel back and forth and producing oscillating currents in the antenna.

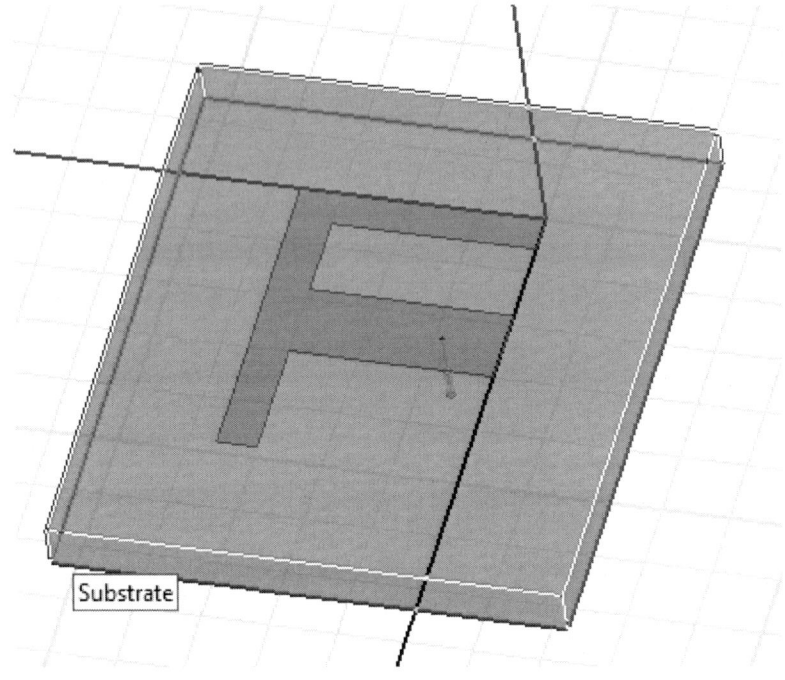

Figure 5. F shaped structure.

Wireless communication, as we all know, is critical. Different antennas are required for wireless communication. There are several types of antennas, such as Yagi Uda, helical, and horn. However, all of these antennas have a high gain and bandwidth, as well as a huge size. As a result, we don't use these antennas for wireless communications. We use tiny strip patch antennas for wireless communication. Micro strip patch antennas have been more popular in recent years because to their inherent benefits of low profile, cheap cost, low weight, circuitry integration, and support for both linear and circular polarisation.

4. Circuit Analysis of MPA

Calculating a microstrip antenna's broadband matching potential necessitates determining the lumped wideband equivalent circuit that simulates the antenna's input impedance across a large frequency range, i.e., from zero to the frequencies that contribute the most to the Bode-Fano integral. By applying the Fano or Youla gain-bandwidth theory to the lumped equivalent circuit of the antenna, the maximum theoretical return loss of a microstrip antenna may be computed vs bandwidth. Although impedance matching networks may increase a microstrip antenna's impedance bandwidth, the actual bandwidth is always less than the theoretical maximum bandwidth provided by the Fano or Youla theories. The physical dimensions of a rectangular microstrip patch antenna (RMPA) have been linked to the narrowband equivalent circuit. Kajfez presented a method for calculating the equivalent circuit of double resonant microstrip antennas using a systematic approach. Kim developed a method for calculating the wideband equivalent circuit of broadband antennas later. In the case of rectangular and wideband E-shaped microstrip antennas, however, the singular value decomposition approach used failed to converge to physically realisable circuits. The analogous circuit's topology is corrected in this work, resolving the problem in the case of rectangular and wideband E-shaped microstrip antennas.

In general, optimization methods are used to find the wideband equivalent circuits of microstrip antennas. Kajfez established a systematic technique to compute the beginning values of the involved optimization in the case of the equivalent circuit that simulates double resonant microstrip antennas. On the one hand, no systematic method for estimating the starting values of the optimization process in the case of wideband equivalent circuits that model antennas in more than two radiation resonances has been provided; on the other hand, as the number of resonances that equivalent circuits of antennas should model increases, the number of parameters of the equivalent circuit increases as well, resulting in increased convergence and computation cost of the involving optimization technique. This issue is also addressed by the suggested solution. Reactively loading may be used to modify the radiation pattern of a rectangular patch antenna. Many articles have been written on different reactive loading strategies for regulating the pattern. The loading position is the sole method for power transmission from the feed to the space if a certain radiation pattern is required. The loading voltages must maintain basically particular magnitude and phase values for a specified radiation pattern from a loading antenna. As a result, designing a loading network to produce a certain radiation is desired. The development of microstrip antennas has attracted a lot of attention in recent years. Others have looked at the consequences of using short circuits, varactor diodes, and slots to load an element. The goal of this research is to see how loading a microstrip element with active reactive load affects its radiation performance. Wide band gain may be accomplished in a single layer microstrip antenna by using the multiresonance features of single stubs, shorting posts, reactively loading, cutting slots, and adding lumped components or active elements. All of these approaches, however, have poor radiation properties, are complicated, and have larger element sizes.

The verification of the theory of loaded microstrip elements, as well as the usage of reactively loaded elements for radiation pattern, are the primary applications of this research. The consequences of loading a microstrip radiator with varied loading points are illustrated both theoretically and via simulation. The antenna's radiation pattern may be synthesised by selecting the suitable characteristic impedance and reactive loads at the antenna's load terminals. In this work, several load configurations are discussed. A negative capacitance contained inside the rectangular patch provides reactive loading, and their best placements for improving radiation performance have been investigated. A microstrip antenna's input impedance is determined by its geometrical form, size, loading circumstances, and feed type. As a result, antenna input impedance is a critical design parameter that affects radiated power and impedance bandwidth. The bandwidth constraints in most applications are caused by an impedance mismatch. Because of their high reactances, microstrip antennas have a small beam width and poor gain in this regard. Several researchers have published reactive loaded antennas for enhancing antenna performance during the last 10 years. Depending on the shape, the radiation efficiency decreases. Because impedance changes are the primary bandwidth limiting issue, the commonly used reactively loaded impedance matching approach is an effective technique.

At a receiver, an antenna turns an electromagnetic signal into an electrical signal, and at a transmitter, an electrical signal into an electromagnetic signal. It also serves as a link between the transmitter lines and the outside world. Because of the significance of wireless communication systems, more work is being put into developing and implementing innovative microstrip architectures, from miniature electronic circuits to antenna arrays. Design of microstrip antenna arrays, which are excellent candidates for adaptive systems in current and future communication systems, is one key use. Light weight, cheap cost, planar or conformal layout, and ability to integrate with electrical or signal processing circuits are their key features. A microstrip patch antenna features a radiating patch on one side and a ground plane on the other side of a dielectric substrate. The patch is usually composed of conductive metals like copper or gold and may be manufactured into any form. On the dielectric substrate, the radiating patch and feed lines are normally photo etched. The patch is usually square, rectangular, circular, triangular, elliptical, or any other common shape to ease analysis and performance prediction. The fringing fields between the patch edge and the ground plane are what cause microstrip patch antennas to emit. A thick dielectric substrate with a low dielectric constant is ideal for improved antenna performance because it gives higher efficiency, broader bandwidth, and better radiation. However, such a setup need a bigger antenna. In order to build a tiny microstrip patch antenna that is efficient, a compromise between its dimensions and performance must be established. Fractal design for the microstrip antenna is achieved by cutting slits on the edge of the stimulated patch surface. The modified Wolff model (MWM), an enhanced version of the cavity model, was reported to properly calculate the resonance frequency of the circular microstrip antenna on the lossless substrate. We use the MWM to compute the resonance frequency, bandwidth, and input impedance of a probe fed circular patch on a thick lossy substrate in this paper. The field theoretic techniques are used to calculate the parameters of the circular patch on thick substrate, and the usual cavity model is not deemed correct.

We gathered experimental findings for substrate thicknesses ranging from 0.003lg to 0.11g from seven published sources. In a dielectric medium, the lg is the guided wavelength. These data were used to compare the computed results obtained by the current MWM and three commercial software, namely the MOM based Ensemble, the standard cavity model used in the PCAAD by Pozar, and the multiport cavity model (MCM) used in the Micropatch by Benalla et al., to the common experimental results. In every example, MWM's findings are substantially closer to the experimental results than the commercial software's calculated results. The modified Wolff model (MWM), an enhanced version of the cavity model, was reported to properly calculate the resonance frequency of the circular microstrip antenna on the lossless substrate. We use the MWM to compute the resonance frequency, bandwidth, and input impedance of a probe fed circular patch on a thick lossy substrate in this paper. The field theoretic techniques are used to calculate the parameters of the circular patch on thick substrate, and the usual cavity model is not deemed correct. We gathered experimental findings for substrate thicknesses ranging from 0.003lg to 0.11g from seven published sources.

In a dielectric medium, the lg is the guided wavelength. These data were used to compare the computed results obtained by the current MWM and three commercial software, namely the MOM based Ensemble, the standard cavity model used in the PCAAD by Pozar, and the multiport cavity model (MCM) used in the Micropatch by Benalla et al., to the common experimental results. In every example, MWM's findings are substantially closer to the experimental results than the commercial software's calculated results. In recent years, there has been a lot of research and development into microstrip patch antennas. Radars, satellites, broadcasting, and radio frequency identification are just a few of the applications where they're employed. A microstrip antenna, in its most basic form, consists of a radiating patch on one side of a dielectric substrate with a ground plane on the other. Microstrip patch antennas offer greater benefits and better prospects than conventional antennas. They are smaller in size, lighter in weight, and can easily be combined into RF and microwave systems. Microstrip antennas, on the other hand, have several disadvantages, such as restricted bandwidth, low power handling capabilities, low gain, and poor impedance matching. These issues, however, may be progressively solved with technological breakthroughs and substantial study in this field. Previously, circular polarisation was achieved by feeding the antenna to several sites and shifting the phase by 90 degrees. The feeding was done directly (without a slot) using a coaxial cable or micro-strip line at the time. Circular polarisation is generated by making the antenna somewhat rectangular with the feeding in one location (instead of square). Either by chopping off two of its corners or cutting a diagonal hole in its metallization. Sharma et al. investigated these three topologies.

The antenna with truncated corners was the final decision in this project. This design is easy to imagine since it has a degree of freedom smaller than the opening in the metallization, in addition to retaining symmetry on the diagonal. While the truncation is symmetrical, the latter might vary in length and breadth. Sharma claims that the antenna with truncated corners has the lowest axial ratio. However, compared to the other topologies, it has a somewhat lesser bandwidth (axial ratio). Broadband antennas have previously been designed using a variety of approaches. Large bands have been recorded using insulated slots in patches, as well as the inclusion of several slot forms at the radiating element, such as the L-shaped slot, T-shaped slot, H-shaped slot, and fractals slot. A U-shaped slot will be employed on the patch of the coaxially fed square patch antenna as one of these ways.

These antennas also have a wide variety of uses in long-range and wireless identification and communication systems, such as RFID, which is one of the newest identifying methods and where the system's size is largely determined by the antenna's size. A new electrical model is created and compared to a physical patch in order to examine its structure. To resolve the

Maxwell equations and subsequently analyse the performance of the antennas, a variety of mathematical techniques are used. Three of these are often used in simulation software:

1. The method of moments (MoM) is utilised in ADS software, among other things.
2. The programme CST Microwave Studio employs the finite integral approach (FIT).
3. The programme HFSS employs the Finite Element Method (FEM).

The technique of moments will be used to examine the electrical model's performance in this paper, and the findings will be compared to those produced using the Finite Integral and Finite Element Methods.

5. Conclusion

Micro strip patch antennas, on the other hand, have poor gain, limited power handling, restricted bandwidth, and low efficiency. To address these limitations, we use various antenna forms such as E-shape, C-shape, H-shape, S-shape, and so on. S-shaped antennas might be utilised to offer extra capabilities such as a broader instantaneous frequency range, larger volumes, and more ideal side lobe distribution in radiation patterns. Two slots are inserted into a rotated square patch to create a micro strip S-shape patch antenna. An antenna's bandwidth is increased as a result of this. By incorporating a PBG structure, the bandwidth is also boosted. This technology produces an antenna with a low volume and low profile configuration, which is readily attached, has a cheap manufacturing cost, and is light in weight. This antenna is designed for use in the C-band. These high data rates are attainable by expanding available bandwidth, which is only possible beyond 300GHz since it is unallocated. As a consequence, the THz band, with frequencies spanning from 300GHz to 30THz, has been investigated as a new frequency spectrum for space communication. THz waves can see through dust, clouds, and other obstructions, allowing for improved space communication in inclement weather. The use of THz frequencies for space communication provides certain additional benefits, such as alleviating spectrum shortages, bandwidth restrictions, and interference issues that now plague wireless communications. Because THz waves are extremely directed and disperse less than microwaves, they allow more secure communications than microwaves. THz wireless communication necessitates compact antennas, decreasing the size and weight of spacecraft and lowering their operating costs.

Chapter 4

Recent Trends in Metamaterials, Challenges and Opportunities

Abstract

The metamaterials under discussion are made up of a regular pattern of rods or a combination of rods and rings. The S-parameters of these metamaterials in a waveguide are investigated and compared to their plasma or resonant structure equivalents. The analytic approach is used to optimize far field radiation, which is then numerically simulated. It has been shown that the metamaterial improves directivity. Simulations of the radiation of a dipole antenna embedded in metamaterial substrates are carried out using commercial software. The technology to develop and execute such structures was offered by recent breakthroughs in Left-Handed Metamaterials. Optimization of structure is feasible by combining analytic techniques for assessing radiation for homogeneous anisotropic slab. As a result, LHM technology has been adopted and improved for the creation of metamaterial substrates. Not only in scattering reflection/transmission phenomena, but also in embedded radiation, metamaterials may be modeled as anisotropic homogeneous materials. We also demonstrate that numerical simulation can solve certain design challenges that are impossible to solve using just analytic methods.

1. Introduction

Electromagnetic metamaterial interacts with electromagnetic waves that are smaller in wavelength than its structural characteristics. Its characteristics must be significantly smaller than the wavelength in order to behave as a consistent material exactly defined by an effective refractive index. The characteristics of microwave radiation are on the order of millimetres. Microwave frequency metamaterials are often made up of arrays of

In: Recent Trends in Microstrip Antennas for Wireless Applications
Editors: S. Kannadhasan and R. Nagarajan
ISBN: 978-1-68507-744-0
© 2022 Nova Science Publishers, Inc.

electrically conductive components (such as wire loops) with appropriate inductive and capacitive properties. The split-ring resonator is used in one microwave metamaterial. At the nanometre scale, photonic metamaterials modify light at optical frequencies. At visible wavelengths, sub wavelength structures have only yielded a few, dubious findings. Metamaterials are one of the most important and fascinating topics in the recent history of Electromagnetic Field Theory. Many scientific studies have been conducted and studied in this subject, with the conclusion that metamaterials offer unique and amazing qualities that may be exploited to improve the main performance of conventional electronics.

Metamaterials are defined as materials that exhibit different behaviour under the influence of electric and magnetic fields than naturally generated materials. The responsiveness of metamaterials-inspired devices at high frequencies is much superior to that of traditional materials-based devices. Metamaterials are used to improve the gain profile, efficiency, and bandwidth needs of different antenna designs and fabrications. We shall examine the history and kinds of metamaterials, as well as their structure-based characteristics and techniques, in this study. We have all heard of a sensor gadget that detects and reacts to input from the physical world. Light, heat, motion, moisture, pressure, or any of a variety of other environmental phenomena might constitute the particular input. The output is usually a signal that is transformed to a human-readable display at the sensor site or sent across a network for reading or additional processing. Significant progress has recently been made in material characterization utilizing properties approaches. Some of these procedures are described for measuring the actual component of the dielectric constant; nevertheless, each approach is appropriate and helpful for certain frequency ranges and materials. Material characterization is used in a variety of fields, including bio-sensing, food quality control, and substrate qualities. Metamaterial antennas are a kind of antenna that makes use of metamaterial to improve the performance of tiny (electrically) antenna systems. The goal of these antennas, like any other electromagnetic antenna, is to propel energy into space. Since of its innovative construction, a metamaterial antenna seems to be much bigger than it is because it stores and re-radiates energy.

A wideband horn antenna laden with metamaterial-inspired particles may be used to conduct various signal manipulations, such as band-pass and band-stop filtering, and polarization modification, on broadcast and received signals. As a result, the modules may be thought of as a novel family of microwave components that execute signal modification directly within a communication system's radiating element. We may minimize the amount of components that make up a communication system by employing these small modules, resulting in significant cost, weight, complexity, and space savings. The purpose of a metamaterial absorber is to effectively absorb electromagnetic radiation like light.

2. Metamaterials

Metamaterial is a breakthrough in the field of materials research. As a result, metamaterials intended to be absorbers have advantages over traditional absorbers, such as greater miniaturization, flexibility, and efficacy. Emitters, sensors, spatial light modulators, infrared camouflage, wireless communication, and usage in solar photovoltaic and thermophotovoltaics are all potential uses for the metamaterial absorber. A device made of a patterned square metallic patch and a metallic ground plane separated by a dielectric layer is meant to expand the bandwidth of a metamaterial absorber.

Different approaches and methods, such as defective ground structure (DSG), photonic bandgap structures (PBG), frequency selective surfaces (FSS), and others, were utilized before metamaterials to enhance the performance characteristics or decrease the mass and volume of microwave passive devices. The features of metamaterials and their behaviour towards electromagnetic RF waves, on the other hand, present a novel method based on the notion of artificial effective media, which is the metamaterials' applicative portion. These artificial structures are made up of unit cells, just as matter is made up of atoms. Because the unit cells are often less than one tenth of the propagation wavelength, or the same size as atoms, metamaterials represent the next level of structural organization of matter is shown in Figure 1. Metamaterials may therefore be thought of as a continuous medium with effective properties, such as effective dielectric permittivity and effective magnetic permeability. The effective parameters of metamaterials may be made arbitrarily small or large, or even negative, by carefully selecting the kind and geometrical arrangement of constituent unit cells.

Material (the subclass of metamaterials with both the effective parameters, i.e., permittivity and permeability, negative i.e., (0) and (0) in a specific frequency range), Single Negative Material (the subclass of metamaterials with either permittivity or permeability negative i.e., (0) or (0) in a specific frequency range), Mu Negative Material (the subclass of metamaterials with both the effective parameters. The design of metamaterials is based on the value of permittivity and permeability. The effective electromagnetic behaviour and response of the metamaterial varies as these two effective parameters change, resulting in different categories of metamaterials such as electromagnetic metamaterials, chiral metamaterials (arrays of dielectric gammadions), and planar metallic on a substrate. When a linearly polarized light is incident on the array, it becomes elliptically polarized when it interacts with gammadions of the same handedness. The TL theory is a valuable tool for analyzing and designing traditional right-handed, RH materials. The core notion behind the TL method to metamaterial design is that employing a dual notion, conventional TL theory may be utilized to evaluate and construct LH metamaterials. An analogous circuit that models a dual transmission line is the inverse of the circuit that models a conventional transmission line. The capacitors are linked in series in the dual case, while the inductors are linked in a shunt arrangement.

Figure 1. Metamaterial.

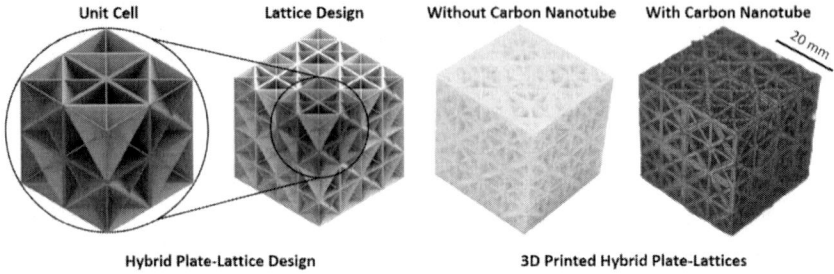

Figure 2. Various shapes of metamaterials.

If the unit cells are tiny enough (far smaller than the wavelength of the propagating signal), the structure may be considered homogeneous, allowing effective permittivity and permeability to be estimated. In a specific frequency range, it has been shown that a dual transmission line has negative effective permittivity and permeability, and hence acts as an LH transmission line. The LH TL is clearly a high-pass filter, while the RH TL is a low-pass filter. Although homogeneous CRLH transmission lines do not occur in nature, they may be built by cascading a number of CRLH unit cells made up of lumped components. Couplers, zeroth order resonators, planar lenses, leaky wave antennas, and other innovative devices have been suggested using this methodology is shown in Figure 2. When the responses of various devices were compared, it was discovered that this method achieved tiny dimensions because to the use of LH metamaterial, but also resulted in quite significant Insertion losses. This flaw has been solved by ForeS unit cells and S-spiral unit cells, which are super-compact LH unit cells. This suggested structure has a smaller insertion loss and a greater quality factor at the resonant frequency.

The optical microscope was created out of curiosity and a desire to examine the tiny world. The optical microscope allowed researchers to detect microscopic features that were previously undetectable to the human eye. With the advancement of the optical microscope, modern biology and medical research have grown, with substantial sections depending on the observation of micro-objects such as cells and bacteria. Optical microscopes have become increasingly practical and vital in recent decades because to fluorescent materials, but there is still a basic concern of how to attain spatial resolution below the diffraction limit. Optical metamaterials are photonic metamaterials, which are electromagnetic metamaterials that are meant to interact with optical frequencies. The source is radiated at optical wavelengths via photonic metamaterials. Furthermore, the photonic metamaterials are distinguished from photonic band gap structures by the sub wavelength period. This is because the optical characteristics are derived from a sub wavelength interaction with the light spectrum rather than photonic band gaps. The current field of study in optical materials is metamaterials with zero index of refraction and negative values for index of refraction.

Negative permittivity or permeability characterize single negative metamaterials (SNG). Another kind of DNG metamaterial is created by combining two SNG layers into one. The slabs of MNG and ENG materials were connected to perform wave reflection studies. Due to their dispersive nature, SNGs, like DNG metamaterials, vary their properties such as refraction index n, permittivity, and permeability with changes in frequency. Metamaterials having negative permittivity and permeability, as well as a negative index of refraction, are known as double negative metamaterials (DNG). Negative index metamaterials are another name for them (NIM). DNGs are also known as left-handed media, negative-refractive-index media, and "backward-wave media."

3. Terahertz Metamaterials

Terahertz metamaterials are a group of artificial materials that interact at terahertz (THz) frequencies and are currently in the works. Passive materials are metamaterials that have negative permeability values and may produce a desired magnetic response. As a result, "tuning" is accomplished by creating a new material with slightly different dimensions in order to generate a new reaction. Terahertz waves range from just before the microwave band's commencement to the far end of the infrared band. Metamaterials is a new topic of study that has the potential to be a very intriguing research subject. Metamaterials are attracting researchers from a variety of areas due to their unusual electromagnetic characteristics. A brief history of metamaterials, as well as several key characteristics, kinds, applications, and modelling methodologies for metamaterials, are described in this work. Metamaterials have resulted in unexpected improvements in electromagnetic response functions, which might open up new opportunities for future device design, component design, and metamaterial features.

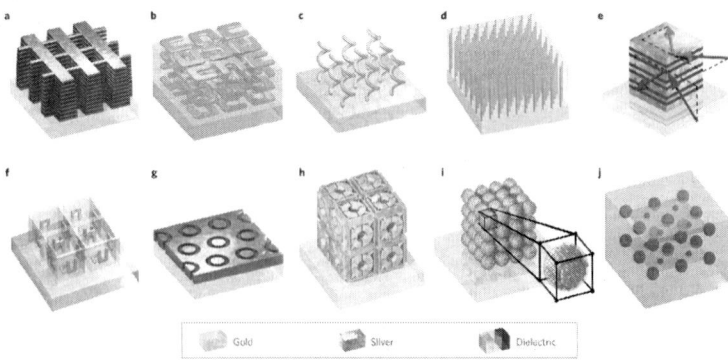

Figure 3. Metamaterials characteristics related with permittivity and permeability.

Spectroscopy, satellite applications, telecommunication, radio astronomy, and other industries employ millimeter and sub millimeter wave equipment nowadays. Despite this, researchers and developers are faced with a new demand and challenge: much reduced weight, improved and well-controlled output characteristics of microwave devices. Unit cells are a kind of metamaterial that consists of a broad range of composite structures made up of manufactured inclusions. Inclusions come in a variety of shapes and are implanted in the base media, which is usually a dielectric substrate. Metamaterials have extraordinary and even counterintuitive properties that are difficult to accomplish technologically and nearly unnoticeable in natural materials. They were able to do so because to the qualities of the basic substrate and carefully chosen unit cell characteristics. Individual cell size, form, and shape are among the latter. Most metamaterials' unit cells are substantially smaller in size and period than the working wavelength (a separate class of photonic crystals with structure dimensions that can be equal to wavelength is not take into account currently). As a result, such materials may be modeled as a homogenous medium with effective permittivity and permeability values.

4. Electromagnetic Inteference Metamaterials

Light propagation is controlled via electromagnetic band gap metamaterials. Photonic crystals (PC) or left-handed materials are used to accomplish this (LHM). Both types of electromagnetic waves contain artificial structures that govern and influence their propagation. The metamaterials are classified as single or double negative based on their separate electric and magnetic responses, which are defined by the parameters permittivity and magnetic permeability. In many electromagnetic metamaterials, however, the electric field creates magnetic polarization and the magnetic field creates electrical polarization, resulting in magneto electric coupling. Such media are referred to as bi-isotropic media because they have anisotropic magneto-electric interaction. They are also referred to as bi-anisotropic media.

Due to the requirement for early detection of illnesses and continuous monitoring of physiological data, the use of wireless telemetry systems in medicine has expanded dramatically in recent years. Because they facilitate communication between the patient and the base station, microwave antennae and sensors are essential components of modern telemetry systems. A split-ring resonator (SRR) is a kind of metamaterial particle with negative permeability that is often utilized in biological sensors. One of the most important uses of the metamaterial antenna is a wireless endoscope, which is a capsule-shaped instrument used for gastrointestinal surveillance and/or therapy.

A highly cost-effective device that can locate an anomaly inside the human body with great accuracy may be built by constructing microwave devices and integrating them with structures inspired by metamaterials. A little change in the water content of tissues may cause changes in the permittivity and conductivity values of the tissues, which is the core premise underpinning cancer diagnosis. The water content of malignant cells is much greater than that of normal tissues. As a result, at microwave frequencies, the tumor's permittivity and conductivity vary from those of normal tissue. To detect, the suggested biosensor. A super lens, sometimes known as a perfect lens, is a lens that goes beyond the diffraction limit by using metamaterials. In traditional optical devices or lenses, the diffraction limit is an intrinsic constraint. A sort of lens with a metamaterial that could compensate for wave decay and rebuilt pictures in the near field was suggested in. The first super lens with a negative refractive index was constructed at microwave frequencies in and gave resolution three times greater than the diffraction limitations. The first near-field super lens that surpassed the diffraction limit was created in the year. By heating sick tissue to a temperature that induces cell death, microwaves are employed to ablate or evaporate it is shown in Table 1.

The generator generates microwave radiation, which is transferred into the patient through the antennas. The application of a precisely regulated dosage of heat to the tumor while sparing the surrounding bodily tissue is required for hyperthermia treatment of cancer. The capacity of a metamaterial lens' negative-refractive index (NRI) to concentrate a source's electromagnetic field is its most notable feature. As a result, it may create the proper focal region in the tissue for microwave hyperthermia therapy. Conformal microwave array applicators with low loss left-handed metamaterial lenses have recently been discovered to be beneficial for hyperthermia treatment of big tumors. Microwave hyperthermia may be used in this manner by heating a big tumor region using many microwave sources at the same time. Strain sensors based on RF-microelectromechanical systems (MEMS) take use of recent metamaterial advancements. It has created a wireless strain sensor to track the healing of long-bone fractures. Under no load, the SRR-based sensor has a particular resonance frequency.

As a result, the course of fracture healing may be properly tracked by measuring the amount of operating frequency change under an applied stress.

We are focusing on merging a natural metamaterial-like structure, the bone, into the construction of a metamaterial, dubbed "Bio metamaterial," based on all of these metamaterial criteria. The symmetrical, repeated building structure of bone possesses several characteristics of a metamaterial. We intend to create a metamaterial with qualities comparable to bone, but stiffer and tougher, by converting the bone structure into a repeating, symmetrical structure is shown in Figure 4. Based on the metamaterial structure chosen, we may achieve a surface feature important for the prosthesis, such as bacterial resistance, high friction, and low wear, without the requirement for coating.

Table 1. Number of papers published in metamaterials

Year	IEEE	IET	Number of Papers Published		WILEY	ACM
			ELSEVIER	SPRINGER		
2015	29	15	10	18	10	28
2016	30	10	20	12	15	29
2017	35	15	20	18	20	30
2018	34	20	10	20	25	22
2019	38	25	10	18	22	28
2020	40	20	20	22	30	22

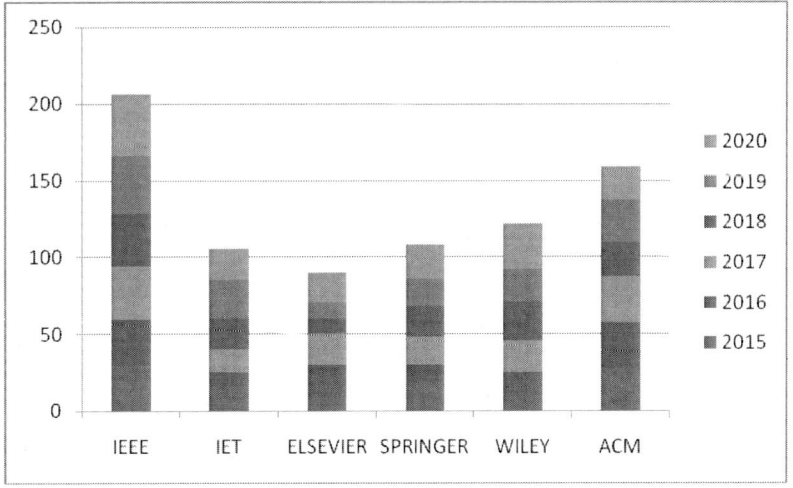

Figure 4. Number of papers published in metamaterials.

5. Conclusion

Recently, there has been an increase in the use of innovative ways to develop microwave components and parts. One of them is using metamaterials, a novel class of materials with unique characteristics, into microwave device building. The development of improved microwave devices is influenced by the development of manufacturing and improvement structures with regulated electromagnetic characteristics. The number of papers dedicated to artificial materials exhibiting novel physical phenomena is steadily increasing. As a result, materials and media with peculiar electric and magnetic characteristics are now attracting a lot of scientific attention. Engineers and physicists are increasingly needed to focus on and create technology especially for healthcare. In the last decade, metamaterials has been one of the most active fields of study with possible applications in healthcare. The broad usage of metamaterials in medical applications is summarized in this publication.

Chapter 5

5G and 4G Communication for Microstrip Patch Antennas

Abstract

People's daily lives and social activities have been significantly impacted by the ongoing development of data traffic and gadget connection. The number of mobile networking devices is predicted to reach 100 billion by 2020, because to very rapid development in wireless consumer items and ongoing expansion in the Internet of Things. Mobile data traffic is increasing (at least) every year, and by 2019, the demand for multimedia and other data-hungry smartphone apps will outnumber wired traffic. Major research institutes and WiFi providers are starting to create the next generation system, namely the fifth-generation (5G) of wireless networks, although the majority of people are getting 10 megabits (Mb/s) of streaming powered by 4G-mobile.

1. Introduction

5G will be able to handle data speeds of more than 1 gigabit per second, ideally around 10 gigabits per second. Power efficiency, minimal latency, and a considerably larger number of smart devices linked are all key benefits. Millimeter wave (mm-wave) frequencies will be used in future 5G cell architectures. 5G is expected to provide a broad range of multi-Gigabit-per-second (Gbps) data speeds for a variety of interchanges. Mixed media, live 3D video, high-end cloud data, and intelligent gaming are some of its applications. Antenna design for contemporary mobile phones is a difficult undertaking, according to all reports. Future 5G worldwide standards and the promotion of 5G networks would need antennas that are significantly smaller yet more capable.

In: Recent Trends in Microstrip Antennas for Wireless Applications
Editors: S. Kannadhasan and R. Nagarajan
ISBN: 978-1-68507-744-0
© 2022 Nova Science Publishers, Inc.

Antenna design is determined by the operational frequency and bandwidth needed. The International Telecommunications Union (ITU) produced a list of suggested 24–86 GHz frequencies (24.25–27.5, 31.8–33.4, 37–40.5, and 40.5–42.5 GHz) to align the homogeneity of mm-wave frequencies over the world. As we raise signal frequency, atmospheric attenuation influences RF transmissions. Signal absorption by atmospheric gases like as O2 and H2O is a major source of attenuation in free space. From 45 to 60 GHz, the absorbing impact grows significantly. At sea level, average air absorption of mm-wave displays the lowest attenuation with the maximum frequency in the 5G spectrum range from 26 to 43 GHz. Furthermore, the FCC published a notice of proposed regulations for flexible services in the 28, 37, 39, and 64–71 GHz bands.

As a result, spectrum in the ranges of 26, 28, and 43 GHz will be required for 5G services. Smaller component size, greater bandwidth, harmonic suppression, and undesirable cross-polarization of higher orders are all benefits of adopting DGS components. DGS causes disruptions that impair the ground's homogeneity and the continuity of surface currents. DGS symmetrical structures serve as resonant gaps that are immediately put on each side of a microstrip line to effectively link the feed line. DGS also changes the shield current distribution of a ground defect to allow for controlled excitation and electromagnetic energy passage through the substrate, modifying the capacitive and inductive response of the transmission line. DGS makes a multiband antenna by increasing the effective capacitance of an antenna or filter, resulting in the production of several resonant frequencies. The required resonance frequency may be tweaked by choosing the right shape and putting the antenna in the right spots. Several wideband and multiband antenna topologies have previously been published, with DGS included as symmetry/asymmetry and a single/period structure as slots or apertures in the radiating patch or ground level. The small antenna design and short wavelength at mm-wave frequencies allow for simple integration into smartphones, wireless LAN bridges, and tablets employing a broad selection of tiny antenna components.

Meta-material structures have been used in various investigations for energy harvesting and absorption applications. A wideband meta-material absorber, for example, was suggested in that was unaffected by polarisation or incidence angle. A plus-shaped meta-material structure may also be used to create solar energy absorbers. Material properties may also be determined using a meta-material structure based on a triangular split ring resonator. Pattern reconfigurable low profile antennas have also used mushroom-shaped electromagnetic bandgap constructions. Personal wireless gadgets (smart phones, smart watches, smart glasses, and so on) that operate at diverse frequencies are becoming more common. Navigation, driving assistance, video capturing, and accepting/rejecting incoming calls from a linked mobile phone are just a few of the features available on these devices. Furthermore, these developing technology products are projected to operate with the Internet of Things (IoT) concept at 5G (fifth generation) frequencies.

In the near future, smart glasses with 5G technology are likely to replace smartphones. Because 5G technology and the Internet of Things will lead to more connected devices being utilised, the amount of time spent with smart gadgets is expected to rise. As a result, the amount of time spent using smart glasses is projected to rise. People may be exposed to more electromagnetic radiation as a result of this situation than in the pre-IoT era, which could be harmful to human health if the allowed standard values set by the International Commission on Non-Ionizing Radiation Protection (ICRNIP) are exceeded, as it could cause unwanted heating effects on tissue. As a result, understanding the specific absorption rate (SAR) distributions for various frequencies is critical.

However, to the best of the authors' knowledge, this is the first research to look into SAR distributions for smart glasses applications at frequencies of 2.45/3.6/3.8/4.56/6 GHz. Furthermore, whereas coupling element structures are often used in the literature to create tiny antenna designs that may be applied to glasses, this work used a folding dipole and a defective ground structure. Using two distinct human head models, the SAR distribution of the proposed tri-band antennas radiating at Wi-Fi and 5G frequencies incorporated into the eyewear device was examined in this work. The remainder of the paper is laid out as follows. The prototype of 3D glasses used as the frame of the smart glasses model is created in CST Microwave Studio and then built in the 'feasibility study' part. The SAR distributions in the human head as a result of the proposed antennas integrated in the frame of the 3D glasses were then investigated for scenarios when the antenna is embedded in the frame and utilised alone. International standards are used to assess SAR values.

2. 5G and 4G Communication

5G mobile associations with massive MIMO implementation techniques introduce unprecedented advancements in channel capacity, data rates, latency, efficiency, and energy conservation to meet the rising demand for wireless services, compared to SISO systems that do not require expanding bandwidth or transmitted power. These systems are looking for more functionality, smaller size, better performance, ease of integration with other circuit components, and, most significantly, cheaper development costs. To provide secure wireless channels, exceptional isolation and a low envelope correlation coefficient ECC (uncorrelated) among antenna components are critical. Patch antennas are prone to multi-band operation. The patch is loaded with slots, slits, or shortened pins to do this. The highest impact of loading is obtained when the slots are implanted at the maximal magnetic field position. The MIMO antenna used in this study differs from other common antennas in order to meet the requirements of the current 5G wireless communication system is shown in Figure 1. In this paper, we propose a unique concept for a compacted slotted microstrip MIMO antenna operating at 28/38 GHz, which is suitable for future 5G wireless communication. A rectangular form of two components slotted microstrip antenna is assumed for maximum use of practical handset space.

In terms of channel capacity and data throughput, the next generation of mobile devices (5G) will represent a quantum leap in communication technology. As a result, the worldwide tendency is to move mobile system operation to the millimetre wave (mmW) band. In exchange for design complexity and system expense, antenna arrays, generally grouped by subarrays, are suitable choices for the mmW band to compensate for the substantial path loss in that band. Several major worldwide mobile network operators and academic institutes have begun to investigate technology advancements in the fifth generation of cellular phones (5G). Different millimetre wave (mmW) frequencies and their propagation properties are described in this paper. In comparison to fourth generation (4G) cellular networks, extensive propagation experiments provided in show that mmW systems may offer less interference and higher capacity. Outdoor non-line-of-sight (NLOS) coverage is also feasible up to 200 metres from a low-power base station. mmW propagation is highly feasible using directional, high gain antennas, according to channel tests carried out in.

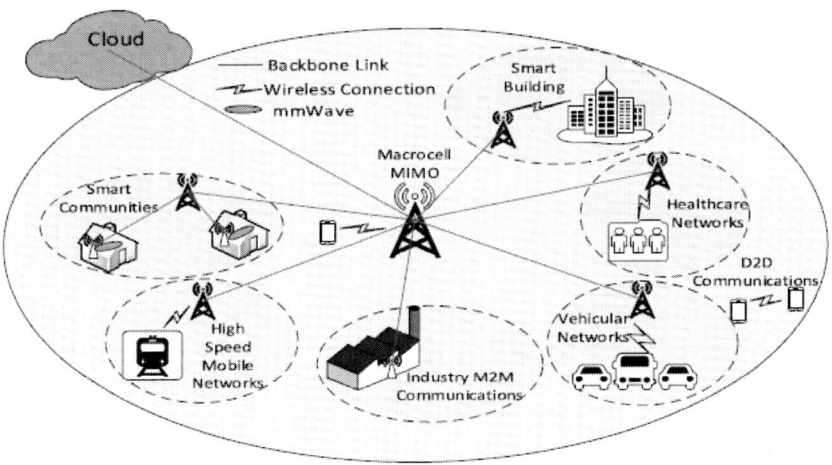

Figure 1. Communication of microstrip patch antenna.

For 5G system simulations and designs in the mmW band, several route loss and channel models, as well as outage probability, signal-processing algorithms, and quality of service aware techniques, are supplied. As previously stated, industrial and academic groups have recommended using the mmW spectrum to expand available capacities and data rates while drastically reducing latency and component size. The Federal Communications Commission (FCC) is considering two mmW Gbps Broadband (MGB) candidates for next-generation Mobile Radio Services, which are 24.25 GHz–24.45 GHz, 25.05 GHz–25.25 GHz, and the Local Multipoint Distribution Service (LMDS) frequency bands operating at frequencies 27.5GHz–28.35 GHz, 29.1GHz–29.25 GHz, and 31GHz–31.3GHz. The primary obstacles in employing these frequencies include susceptibility to blocking, propagation losses, shadowing, large-scale attenuation of materials and human bodies, and air absorption. Highly directional antennas should be used to circumvent these obstacles. The radio energy is focused into a narrow beam with high gain using an antenna array of several elements, which increases the effective isotropic radiated power (EIRP) without increasing the transmitter power.

The growing need for data rates has prompted experts all around the world to create backward-compatible cellular device ecosystems. Microwave engineers have been prompted by the growth of smartphones to build subsystems that are compatible with current protocols. New technologies have emerged, and portable devices with Long-Term Evolution (LTE) for voice and data applications are on the way. LTE technology uses three low-frequency bands to operate: LTE700 (698–787MHz), LTE2300/Class 40 band (2300–2400MHz), and LTE2500/Class 7 band (2500–2690MHz). MIMO (Multiple-Input Multiple-Output) is one of the best ways to get a high data rate. The design challenge with LTE antennas is that they have a significant physical footprint, making integration in mobile terminals difficult. Future transceivers will need to include 5G hardware in addition to 4G LTE antennas. For the construction of 5G antennas, frequencies for mmWave 5G are projected to be in the 28GHz range. It's difficult to provide orthogonal pattern variety with a small physical footprint.

Backward compatibility of future devices will be hampered by the use of 4G LTE and mmWave 5G MIMO antennas on the same module. It will reduce the antenna's available footprint, allowing them to fit inside the handset's restricted area. It's difficult to coexist mmWave 5G antennas with 4G LTE antennas on the same module. The disclosed designs show the co-design of 4G LTE andmmWave 5G antennas, however their effective radiating volume after integration is extremely huge. Conformal antennas have been widely developed for a variety of applications using non-flat surfaces (singly curved or doubly curved). Due to a limitation of space, a compact design is required.

In the mobile device industry, the demand for multifunction devices such as smartphones is increasing. WiFi, high-speed mobile internet (3G and 4G), GPS (Global Positioning System), GSM (Global System Mobile) ou (Global System for Mobile Communication) and Bluetooth® are just a few of the mobile technologies available on these devices. In a globalised society, the speed of the internet has been steadily improving over time in order to do real-time video chats utilising, for example, a mobile operator's connection. Furthermore, the displays of these gadgets are growing in size to provide a better user experience during high-resolution video conversations.

Despite the fact that screen sizes are growing, users are also searching for slimmer gadgets. As a result, in order to minimise the size of a smartphone, its internal components must be shrunk. As a result, several approaches are used, such as the production of integrated circuits, which reduces the circuit size significantly. It is also important to employ printed antennas, which have a very thin thickness. The need for faster mobile internet has fueled research and development of new technologies such as 4G, which can provide data rates of up to 100 Mbps. Despite the fact that this technology has not yet been fully implemented in several areas, there is a slight increase in research and development of the Fifth Generation, or 5G, of mobile telecommunications, which is capable of operating at extremely high data rates and using electromagnetic spectrum frequency bands above 25 GHz. Furthermore, the potential of an increase in the number of devices linked at the same time after the introduction of the Internet of Things (IoT) has worried mobile technology businesses, who are looking for a way to satisfy future demand effectively, emphasising the need of 5G deployment. This study proposes the construction of a microstrip antenna capable of satisfying all desired qualities in a smartphone, including future 5G technology, using just one antenna in order to minimise the mobile device's size.

According to the IEEE standard definition, an antenna is a device that acts as a medium for radiating or receiving radio waves, or the medium of transition between free space and the waveguide. The microstrip antenna was selected for the project because of its exceedingly thin thickness. Furthermore, this antenna must have ultra wideband antenna characteristics. The microstrip antenna was initially employed in space applications in the 1970s (despite their design in the 1950s), making them quite popular in this field at the time. This kind of antenna is now extensively utilised in civil and commercial applications, particularly in mobile devices (such as smartphones). The three levels of a microstrip antenna are as follows: The ground plane is a metallic layer over a substrate atop another metallic layer. This third layer (ground plane) may have a variety of design forms, which are referred to as configurations. These are the settings that, for example, determine the device's operating frequency. The circular and rectangular geometries of this kind of antenna are the most popular since they are simple to project (for a limited number of resonance bands), create, and manufacture. These forms also have appealing radiation properties and low cross polarisation.

These antennas have a low profile (they're incredibly thin), so they'll fit on both flat and non-flat surfaces. They also offer considerable adaptability owing to the ease and economy of their production, which is comparable to the printed circuit board manufacturing method. Furthermore, when fixed in solid surfaces, this antenna is mechanically resistant. Finally, several antenna properties, such as resonance frequency, polarisation, radiation pattern, and impedance, are simple to change. Due to the growing number of internet users, wireless communication technology is continuously evolving. We've seen 1G, 2G, 3G, and, most recently, 4G LTE technology. Lack of viable frequency resources is one of the major problems affecting today's wireless communication. To address this issue, 5G wireless communication research has begun in the millimetre frequency spectrum, which spans 20 GHz to 300 GHz. The frequency spectrum utilised for 5G research is typically between 24 and 60 GHz. With the Internet of Things, 5G technology has been embraced in a variety of areas (IOT). Connecting millions of devices is one of the aims of 5G technology. Smart cities, smart transportation, and robots may all benefit from 5G technology in the future. The rapid shrinking of mobile devices has led to the development of small antennas that can fit within such gadgets without compromising their functionality. Microstrip patch antennas were popular in the twentieth century as a result of this. A microstrip patch antenna is made out of a thin metal foil put on a substrate with a ground plane underneath it. This microstrip patch antenna can be readily fitted on the surface of a PCB and may also be utilised in mobile devices. The microwave and millimetre frequency bands are where these antennas are most often utilised.

3. Challenges

Cellular communication systems' reputation has risen substantially in the previous decade, and market demand continues to rise. Cellular communication will need to improve in the near future in terms of QOS (Quality of service) and execution. Furthermore, without antenna, future technologies such as wireless communication would not be possible. In this regard, antenna design configurations must be advanced in order to meet new demands for the benefit of society. Microstripped patching antennas are widely utilised because to advantages such as light weight, small dimensions, cheap production costs, and the capacity to double and triple frequency bandwidths. The design of a directional microstrip antenna for wireless applications necessitates a high level of access. The small band width of a microstrip antenna is its most essential attribute. Many solutions have been established and produced for the development of micro-strip antennas bandwidth to eliminate these distinctive inhibitions of restricted impedance and axial ratio (AR), capacity of the transmission rate of data. Analysts from all around the world created a variety of forms and design frameworks. Furthermore, as a result of the study of the communication globe, the number of customers is expanding, resulting in a broad variety of deficiencies in the society. Future wireless communication technologies, such as 5G and prior generations, will likely employ millimetre to micrometre frequencies, as specified by the International Teleconference

Union. It is necessary to construct such a microstrip antenna system for instant access to web-services and distant communication at frequencies greater than 6GHz. Microstrip antennas are simple in design, have small measurement dimensions, and vibrate throughout 6GHz bandwidths is shown in Figure 2. With the increasing conservatism of electronic building, at least two narrowband constructions must be installed together. The nano sized strip antenna that is employed at multiple levels of frequencies is difficult to create. Our suggested microstrip antenna will be built using the High Frequency Structure Simulator.

According to a study conducted by Computer Information System Company (CISCO), mobile data traffic might reach 4.8 Zettabytes (ZB) each year by 2022, or 396 Exabytes (EB) per month, up from 1.5 ZB/year or 122 EB/month in 2017. According to a CISCO study, roughly 50 billion smart devices will be connected to the Internet by 2020. IoT has transformed ubiquitous computing in the recent decade as a result of multiple application areas such as smart cities, smart agriculture, and smart health, among others. The Internet of Things (IoT) is a concept that comprises a collection of smart devices and sensor nodes. Sensor nodes keep track of certain parameters and communicate them via the internet. According to a study, billions of gadgets would be in use by 2020, with an average of six to seven devices per person. More than a trillion sensor nodes will be connected to the Internet by 2022. In the next twenty years, it is predicted that around 45 trillion gadgets will be connected to the Internet. It is important to look for a 4G option to deliver continuous services to these mobile devices. It is estimated that every ten years, a new generation of cellular phones will be introduced. The fourth generation of cellular networks, known as 4G, was introduced in 2011, and 5G networks are projected to be standardised and implemented by 2020.

The 5G network's scope extends beyond radio technologies to include fixed host communication, cloud infrastructure, and other services. The 5G mobile network extension services strengthen the ecology of the telecommunication network and deliver energy-efficient services to the healthcare, agricultural, and smart city projects. From personal communication to societal connectedness, 5G lays the groundwork for digitalization. Digitalization creates fantastic opportunities for mobile communication, but it also poses significant obstacles to mobile communication systems. For voice communication, the first generation (1G) of wireless networks was standardised in 1981. It could handle data transmission rates of up to 2.4kbps. Advanced Mobile Phone System (AMPS), Nordic Mobile Phone System (NMTS), Total Access

Communication System (TACS), and other 1G-access technologies were the most popular. In 1G, analogue signals were responsible for carrying voice. It has a number of flaws, including poor signal quality, insufficient capacity, and insecure and unreliable handshakes. off

Wireless networks of the second generation (2G) were standardised in 1990. It was designed mainly for voice communication and was capable of data transmission speeds of up to 64kbps. It was also capable of modest data transfer. Global Systems for Mobile Communications (GSM), Code Division Multiple Access (CDMA), and IS-95 were the most prevalent 2G-access technologies. Text communications, photo messaging, and MMS Multimedia Messaging Services were all possible with 2G technology (MMS). It may also allow secure point-to-point communication, which means that only the intended recipient can see and hear the message. Low data rates, restricted cell capacity, greater handover delay, limited mobility, and other significant concerns plagued 2G. Additionally, 2G capable phones have limited capabilities.

It was a development of second-generation wireless technology. It introduces the General Packet Radio Services (GPRS), a packet-based switching technology (GPRS). It may also offer enhanced communication via the use of packet switching and circuit switching methods, as well as 2G services. It can transport data at a rate of up to 144kbps. GPRS, Code Division Multiple Access-2000 (CDMA2000), and Enhanced Data Rate for GSM Evolution were the most popular 2.5G-access technologies (EDGE).

In the year 2000, the third generation (3G) of wireless networks was standardised. The primary goal of 2G was to provide voice communication and high-speed data transmission of up to 2Mbps. Wideband Code Division Multiple Access (WCDMA), CDMA2000, and Universal Mobile Telecommunications Systems (UMTS) were the most popular 3G-access technologies. To take use of the benefits of 3G smartphones, dedicated programmes for video chatting, online gaming, email, and social networking services like Facebook and Orkut were created.

It was standardised in 2008 as a 3G wireless network expansion. It was created largely to increase the data throughput of current 3G networks, and it can handle data transmission speeds of up to 3.6Mbps. HSDPA (High Speed Downlink Packet Access) and HSUPA (High Speed Uplink Packet Access) were the most common 3G access technologies (High Speed Uplink Packet Access). As an upgraded version of the 3G network, the 3.75G system was suggested. High Speed Packet Access Plus (HSPA+) was the technology employed. Long-Term Evolution (LTE) and Fixed Worldwide

Interoperability for Microwave Access were the technologies employed in it (WIMAX). These technologies may provide high-speed services to several users at the same time, such as on-demand films, composite web services, social networking services, and so on. Although 3G technology represents a significant advancement in the world of communication, it is plagued by high implementation costs, compatibility issues with 2G systems, and strong magnetic wave radiation that harms our brains, among other issues.

Wireless networks of the fourth generation (4G) were standardised in 2010. 4G is intended to manage data transfer speeds of up to 300Mbps while maintaining a high level of service quality (QoS). Users using 4G may stream high-definition (HD) video online and play online games. Voice over LTE (VoLTE) is the most widely used 4G access technology (use IP packets for voice). Long Term Evolution (LTE) is currently being standardised by the 3G Partnership Project (3GPP) (LTE). It delivers safe mobility and minimises latency for key applications. It also allows IoT-enabled devices to communicate efficiently. In terms of hardware and implementation, 4G is more expensive than 3G. High-end multipurpose gadgets that are compatible with 4G technology are required for communication.

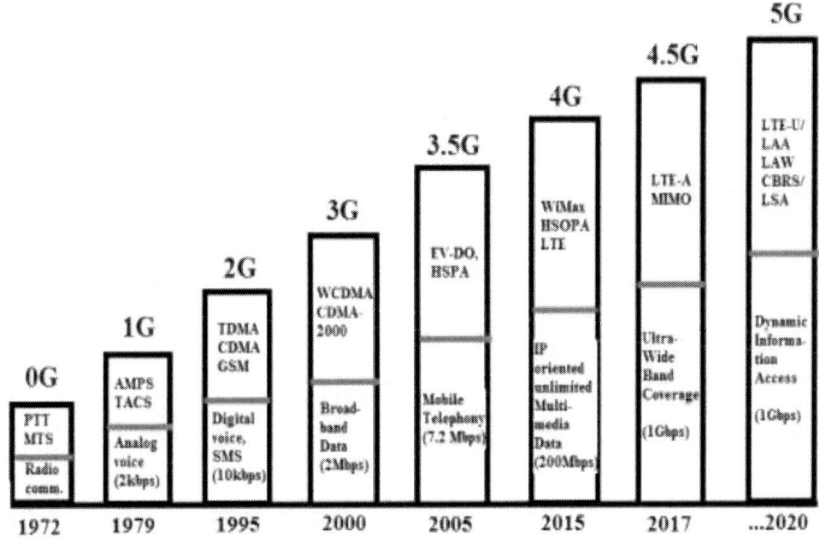

Figure 2. Generation of communication technologies.

The distribution of material to mobile devices is the primary emphasis of 3G and 4G networks, rather than efficient delivery. The 5G wireless network can give services to billions of devices with near-zero latency. In the year 2020, 5G is projected to be standardised. With QoS, 5G can manage data transmission speeds of up to 10Gbps. Higher speeds enable you to view Ultra High Definition (UHD) video online and play online games. Challenges are an unavoidable component of new growth. In compared to 3G and 4G networks, the fundamental goal of 5G is to deliver high-speed mobile broadband and greater throughput, as well as ultra-low latency, excellent dependability, and security. The infrastructure for 5G is complicated. It is necessary to build a high number of Base Stations (BS) inside a limited geographical area. Although it will raise network costs, it will enhance high data transmission rates and minimise energy usage. Cognitive Radio Networks (CRN) and Massive Multiple Input Multiple Output (mMIMO) architecture will be used to achieve high-speed. In comparison to communication devices, mMIMO employs a high number of antennae to boost efficiency. The frequency range for mMIMO is 30-300 GHz, with a wavelength of 1-10 mm. Half-duplex [6] communication is used in 4G networks, which means there are two independent channels for uploading and downloading. 5G, on the other hand, is intended for full duplex connectivity, which means that both access and backhaul will use the same channel. Despite the fact that it will enhance connection capacity, preserve frequency spectrum, and be more cost-effective, actual application will be challenging owing to interference. As a result, a device to neutralise the influence of interference is also required.

The 4G Radio Network (RN) consumes around 70% to 80% of total power. This results in a massive quantity of CO_2 being released into the atmosphere, which has a detrimental influence on the environment. For this, numerous 5G methods have been suggested. To make 5G environmentally friendly, it uses Cloud-Radio Network (CRN), Visual Light Communication (VLC), millimetre wave (mmWave) communication, D2D communication, and Massive Multiple Input Multiple Output (mMIMO) designs. The 4G network's roundtrip latency is roughly 15 milliseconds (ms). It is expected that the 5G network would have exceptionally low latency, resulting in fewer packet loss and improved network stability. To do this, 5G networks may use efficient caching, mmWave, and mMIMO design. The latency of the 5G network will be exceedingly low. It will have a direct impact on service quality, end-to-end delivery, connection, and dependability, among other things. Intelligent equipment, load balancing methods, and delay bound QoS are all used to increase QoS. Small cell network architecture will be used instead of base station (BS) centric architecture, or more specifically device centre architecture, in the 5G network. It's possible that the cell is a microcell or a picocell. The ideal or non-ideal backhaul design connects these cells. Because the cell is smaller, there will be more mobility and handover.

4. Conclusion

Traditional mobile communication networks are focused on providing communication services to individual consumers, while 5G is focused on providing services to individuals as well as businesses. Mobile IoT devices have a lower security need, but high-speed mobile services have a higher security requirement. Denial of Service (DoS) attacks, hijacking attacks, signalling storms, resource (slice) theft, security keys exposure, IMSI catching attacks, IP spoofing, scanning attacks, TCP level attacks, Man-in-the-middle attacks, configuration attacks, penetration attacks, user-identity theft attacks, and so on are some of the major security challenges in 5G networks.

Chapter 6

Wearable Antennas for Microstrip Patch Antennas

Abstract

Portable electronic gadgets, such as cellphones, have become an integral part of daily life, allowing users to do more than simply make phone calls by providing internet access, multimedia, and a personal digital assistant. This kind of "always on" and always connected condition might be seen as a precursor to the ubiquitous computing concept. According to the report, substantial global research has been conducted on new wearable gadgets, with wearable computers potential of serving as forerunners to smart clothing. In this way, smart clothing will allow wearable electronics to break free from the limits of the rigid box and fuse with textile technology. The inclusion of antennas into uniforms will have the added advantage of removing bulky gadgets that may become entangled. As technology advances, new and diverse applications such as wireless transactions, general network connections, navigation support, location-based services, tourism, security, emergency, child protection, intelligent transportation systems, military applications, smart suits, backpack radar, battlefield personnel care, medical monitoring, smart diagnosis, ageing care, biosensors, space applications, and astronaut monitoring have emerged.

1. Introduction

Wearable electronics, on the other hand, must be lightweight, flexible, compact in size, affordable, able to endure harm from obstacles (robust), and pleasant to wear. With the progress of communication technology, UWB antennas have attracted more and more attention since the Federal Communications Commission (FCC) permitted the commercial use of frequency bands from 3.1 to 10.6 GHz for Ultrawideband (UWB) systems in

In: Recent Trends in Microstrip Antennas for Wireless Applications
Editors: S. Kannadhasan and R. Nagarajan
ISBN: 978-1-68507-744-0
© 2022 Nova Science Publishers, Inc.

2002. The appealing features of UWB antennas, such as low profile, low cost, and radiation properties, mean that they don't need to transmit a high-power signal to the receiver, can have a longer battery life, and their compact size has increased the possibilities of shrinking the size of wearable devices and thus making the fabrication process easier.

Wearable communications systems have been a hot issue in academic circles since 1997. Wearable antennas and systems have been the subject of several articles on design, manufacturing, and applications. Some of these advancements are highlighted in the following sections. Several studies have made significant advances in the use of textile materials as antenna substrates. Flexible metal patches on textile substrates have been produced as wearable antennas. A dual-band wearable antenna was thoroughly explored. Another research on the use of electrotextile materials in the design of microstrip patch antennas and UWB antennas was also published. UWB sensors and their usefulness for medical applications were also taken into account. With all of these results, the market for innovative materials for UWB antenna designs and applications has opened up to a broad variety of prospective needs and research. In light of all of the above, the authors of the present work set out to take steps closer to true wearability. Furthermore, complete success may be reached only when the antenna and all associated components are completely transformed to 100% textile materials, with integrated textile components allowing the electronic suit to be washed and reused. A completely textile UWB antenna is provided in this paper, along with a comprehensive description of the idea, simulation, and manufacturing method. For the conducting portions, two kinds of conducting materials were employed, while the antenna substrate was made of a nonconducting fabric. Materials utilised to create the wearable antenna design, subsequent design procedures, and a set of comparison findings for the suggested design will be described in the following sections is shown in Figure 1. Furthermore, the impacts on the return loss will be addressed in further detail using measurements for each manufactured antenna prototype under bent and totally wet situations. Textile materials are attractive substrates for flexible antennas because fabric antennas may be readily incorporated into clothing.

Textile materials have a low dielectric constant, which decreases surface wave losses and increases the impedance bandwidth of the antenna. Our research focuses on the usage of flannel cloth as a substrate material. Flannel fabric is a kind of 100% cotton fabric that has a smooth, firm, and fluffy surface, making it perfect for wearable applications. Fabric substrate material properties may also serve in keeping the distance between the radiating patch and the ground plane consistent in order to manage the antenna's electrical characteristics. In reality, the fabric substrate material's smooth and firm surface is essential for conductive sheets to be fixed uniformly and firmly on the fabric surface. The resonance frequency of the proposed antenna changes if the copper sheet detaches just from one corner and the spacing between the metal layers changes. Although the thickness of this fabric is almost 1 mm, it is necessary to know its relative permittivity in order to precisely describe the influence of textile materials. The observed relative permittivity of flannel fabric encompassing the frequency range from 300MHz to 20 GHz using the commercial dielectric probe (reflected technique) is roughly 1.7, whereas the loss tangent is about 0.025.

The demand for flexible antennas has skyrocketed during the past year. The improvement in computer architecture and numerical calculation is a key element in the need for such antennas. Microstrip patch antennas are used in satellite communication, telephony, and navigation systems, among other things. The antenna's performance is determined by the substrate's impedance bandwidth, gain, and radiation pattern. Many dielectric materials, including as paper, expanded polystyrene foam (EPS), leather, and others, are available in the market for flexible and wearable applications. These materials are low in relative permittivity, light in weight, and have a low cost profile. Foam is employed as a sub-strate in the suggested antenna design, with a relative permittivity of 1 that is about equivalent to the air dielectric constant. As a result, losses and attenuation are much lower than with other substrate antennas. It increases the antenna's bandwidth.

In the suggested antenna, sand-wiched between two layers of 0.038-width copper tape is employed as a substrate material. The ground is the first layer, and the patch form is the second. The sub-strate height h is 2mm, the dielectric constant is 1, and the operating frequency is 4 GHz. The simulation in this article is carried out with the help of CST software studio, and the results are presented in terms of reflection coefficient, gain, bandwidth, and efficiency.

The use of textile materials as antenna substrates, ground, or patches has advanced antenna technology for manmachine interfaces by leaps and bounds. This will provide designers total creative control over body-worn antenna systems incorporated in smart clothing in the future. Smart clothing may soon become a part of our daily lives. They'll show up in sports uniforms, first-responder uniforms, military, medical, and space applications. One of the most important features of smart clothing is the capacity to communicate wirelessly.

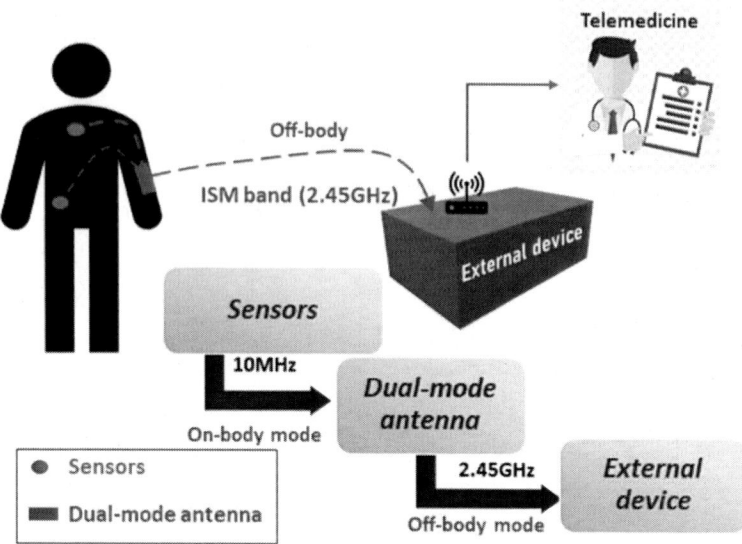

Figure 1. Wearable antenna.

Furthermore, the majority of wearable antennas produced in recent study only work in the 2.45 GHz frequency spectrum for wireless communications. As a result, by enhancing the antenna to resonate at dual bands, multiple mobile network connections may be made at the same time. Due to a high level of interest in Wireless Local Area Network (WLAN) applications for body worn devices, the textile antenna presented in this research is intended to function at 2.4 GHz and 5.8 GHz. According to the Malaysian Communication and Multimedia Commission (MCMC), the cheap cost of equipment and convenience of setting up such networks compared to other networks make it an appealing and hassle-free method of offering short-range wireless communication to the general population. Several viable techniques, such as slot loading, slit, multilayer patch, and frequency reconfigurable by capacitor and shorting pins, are known in the literature and may be applied to enable a single radiating antenna to resonate on dual frequencies. This research focuses on the slot loading approach and, as a result, examines the antenna performance of various types of slots in patches.

Because of the nature of wearable wireless technology, antennas must be flexible, light weight, tiny, and low profile. At the same time, these antennas must be physically strong, efficient, and have a broad bandwidth in order to provide desired radiation properties and user comfort. As a result, the most beneficial choice is the microstrip patch antenna, which can be easily fitted into garments. This project's planned antenna is entirely built of textiles. When compared to a standard micro strip antenna, a textile antenna is unique. The textile antennas in wearable applications are flexible and have a planar structure, so they do not influence the user's wearing comfort. Parametric research on different slot structures have shown that the H- and L-slots perform better than the other shapes. A rectangle shaped patch with a 50 transmission feed line makes up the proposed H- and L-slot wearable textile antenna. In terms of reflection coefficient, S11 under intended reflection coefficient, S11 of below -10 dB between the desired WLAN 2.4 GHz and WLAN 5.8 GHz for both antennas, analysis of the presence and absence of slot is carried out in simulation. Bending angles of 20o, 30o, 40o, 50o, 60o, 70o, and 80o are also investigated further.

2. Wearable Antenna

Textiles have been used in electrical engineering since 1993 (e.g., e-textiles or smart clothes). Wong looked at making radar absorbers out of conductive polymer composites. Wong researched a novel smart material based on this conducting polymer composite material and observed DC and MF electrical characteristics a few years later. Computer systems have become substantially lighter and more compact as the semiconductor industry and communication technologies have developed rapidly. The ability to incorporate electronic components in clothes has been enabled by the compacting of electronic components, making smart clothing a feasible garment for daily usage. Conducting and non-conducting fabrics are increasingly being used in smart apparel designs.

Wearable technology has been steadily increasing since the 1970s, and the popularity of fitness trackers and smart watches has boosted the industry significantly in recent years. Many wearables are used to gather and communicate health-related data with other devices. Instead of attaching wearables to the body or garment, it may be more practical to incorporate them directly into clothes (also known as smart clothing or e-clothing). Wearable technologies, on the other hand, have yet to be completely incorporated into clothes because to the hard materials used in conventional antennas and circuits, making garment integration difficult. Due to their low mass, physical flexibility, and ability to be sewed for garment construction, e-textile antennae and circuits, which are composed of conductive textiles, have a lot of promise for incorporating wearables into clothes.

Wearable antennas should have a low profile, have light weight, have a compact volume, and have a low manufacturing cost. Because of their low bulk, physical flexibility, and simplicity of integration with clothing, e-textile antennas have a lot of promise for wearable antenna design. Wearable antennas need low surface resistivity e-textiles. E-textiles with a surface resistance of less than 0.05 Ohm/Square may match standard copper antennas in terms of electrical performance. Many textile antennas have been developed to function in various frequency bands and on various parts of the human body for various uses. They're made to function in a variety of frequency bands for a variety of applications, including cellular communication, digital television, WFI, Bluetooth, and so on. Aside from antenna design, other studies have looked at how environmental changes or wearing circumstances, such as shape distortion and moisture influence, affect textile antenna performance. For example, researchers discovered that raising the moisture content of a textile antenna increased the material's permittivity and loss tangent, shifting the antenna resonance frequency and lowering antenna efficiency.

The development of textile-based antennas heralds a new era in the use of non-invasive sensors in clothes for real-time health monitoring. Because it covers a broad variety of medical and consumer electronics (CE) applications, the IEEE 802.15.6 group's development of Wireless Body Area Networks (WBAN) in 2010 represented a significant leap in health care for medical uses. As a result, WBAN allows patients to be monitored in real time while going about their daily lives and participating in other activities such as sports. For WBAN applications, IEEE 802.15.6 provides numerous frequency ranges. The Industrial Scientific Medical Band (ISM: 2.4 GHz and 5.8 GHz) stands out among them all. WBAN sensors should have a low power consumption, a low profile, excellent compactness, and be simple to incorporate into textiles. They should also minimise any body interference with the antenna's qualities as much as feasible. The creation of appropriate wearable electronics and antennas to be incorporated into textiles has been realised based on the physical features of textile materials reported in recent research publications. These wireless integrable sensors may be placed within, on, or around the human body. It is feasible to satisfy a demand in personal healthcare systems based on the spread of these wearable sensors: human body communication to gather medical data (on-body) and human-to-human body communication to exchange data with outside networks (off-body). The suggested health monitoring system is outlined in the following paragraphs. The U-shaped antenna is one of the most appropriate antennas for human body communications.

Following the fast growth of wireless communication systems, flexible antennas have attracted a lot of interest in recent years. Flexible antennas are gaining popularity among researchers because to their great flexibility, light weight, cheap cost, simplicity of mounting on a conformal surface, and ease of manufacture. Flexible antennas have been thoroughly researched and constructed for a variety of applications, including biomedical applications, wireless local area networks (WLAN), satellite communication, and vehicle navigation and communication. Innovative designs must meet the criteria of low weight, thin, tiny, bendable, multiband, and green material to keep up with the newest trends in flexible antennas.

Thin and flexible substrates such as composites and polymers will replace traditional rigid substrates like as FR4, Rogers, and Taconic to achieve antenna flexibility. Basalt fibre composite, the flexible substrate in this article, is a green and environmentally beneficial product that does not contaminate the environment. Basalt fibre is a natural substance created by melting volcanic stone at temperatures between 1450 and 1500 degrees Fahrenheit. It's known as volcanic rock silk, and it's an organic continuous thread made by brushing Platinum rhodium alloy. Basalt fibre stood out from the others because of its high tensile strength, making it suitable for use in composites. It also has a strong corrosive resistance and chemical stability, allowing it to endure very acidic, alkaline, or humid environments. Basalt fibre is also electrically insulating, thermally stable, and capable of producing strong composites due to its compatibility with other materials.

Coplanar waveguide (CPW) is the feeding mechanism for the developed flexible antenna since it is ideal for WLAN and wearable applications. Because CPW may be used ungrounded, it has a broader bandwidth and requires just a single layer of radiating element. It may also provide low dispersion, low radiation leakage, and regulate the characteristic impedance, among other things. The suggested flexible antenna is suitable for WLAN systems operating at 2.4 GHz and 5 GHz, with frequencies of 2.4-2.484 GHz for IEEE 802.11b/g and 5.725-5.825 GHz for IEEE 802.11a. The goal of this design is to create a dual band antenna for WLAN systems using spiral strips and C-shaped strips while demonstrating the flexibility and performance of flexible antennas in flat and bending conditions using simulation results.

3. Conclusion

Most governments' major objective in the previous decade has been to minimise healthcare spending owing to an increase in the number of old people. Indeed, according to the US Bureau of Census, the number of senior persons in the United States will have quadrupled from 35 to 70 million by 2025. According to Eurostats, about a fifth of the EU population was over 65 in 2017, and this number is expected to rise to around 29% by 2050. The majority of research has centred on the development of new user-friendly gadgets comprised of smart textile materials.

References

A. A. A. Abdelrehim and H. Ghafouri-Shiraz, "Performance Improvement of Patch Antenna Using Circular Split Ring Resonators and Thin Wires Employing Metamaterials Lens," *Progress In Electromagnetics Research B*, Vol. 69, 137-155, 2016.

Abdelrehim, Amal & Ghafouri-Shiraz, Hooshang. (2016). "High performance terahertz antennas based on split ring resonator and thin wire metamaterial structures". *Microwave and Optical Technology Letters*. 58. 382-389. 10.1002/mop.29580.

Amit Kumar, Jaspreet Kaur, Rajinder Singh. 2013. Performance analysis of different feeding techniques. International Journal of Emerging Technology and Advanced Engineering (IJEATE) conference volume 3.

Arvind Kumar, Mithilesh Kumar.2014. Gain enhancement in a novel square microstrip patch antenna using hybrid structures. International Conference on Signal Processing and Integrated Network.

B. Langat, "Design and simulation of a modified circular microstrip patch antenna with enhanced bandwidth," in Scientific Conference Proceedings, 2014

B. T. Salokhe and S. N. Mali, "Analysis of Substrate Material Variation on Circular, Rectangular and Non Linear Microstrip Patch Antenna," *International Journal of Current Engineering and Technology*, Vol.4, No.3 (June 2014).

B.-L. Ooi, S. Qin, and M.-S. Leong.2002. Novel design of broadband stacked patch antenna. IEEE Trans. *Antennas Propag.*, vol. 50, no. 10, pp. 1391–1395.

Bimal Garg, Neeraj Sharma, " Analysis and design of left handed metamaterial to ameliorate the bandwidth and return loss using CST", *CREST Journals*, Vol 01, Issue 03, pp. 73-79, May 2013

Bimal Garg, P. K. Singhal, Nitin Agrawal, "A High Gain Rectangular Microstrip Patch antenna using "Different C Patterns" Metamaterial design in L-Band", *Advanced Computational Techniques in Electromagnetics*, Vol 2012, pp. 1-5, 2012.

C. A. Balanis, Antenna theory: analysis and design. John Wiley &Sons, 2005, vol. 1.

C. Caloz and T. Itoh, Electromagnetic Metamaterials: Transmission Line Theory and Microwave Applications. New York: Wiley, 2004.

C. R. Simovski, P. A. Belov, S. He, "Backward wave region and negative material parameters of a structure formed by lattices of wires and split-ring resonators." IEEE Trans. *Antennas Propagat.*, 51(10): 2582-2591, 2003

Caloz, C., Itoh, T. Electromagnetic Metamaterials: Transmission Line Theory and Microwave Applications. New York: Wiley-IEEE Press, 2006.

Christophe Caloz and Tatsuo Ioth, "Electromagnetic Metamaterials: Transmission line Theory and Microwave Applications, The Engineering Approach" a John Wiley & Sons INC Publication, 2006

D. M. Pozar, "A Microstrip Antenna Aperture Coupled to a Microstrip Line", Electronics Letters, Vol. 21, pp.49-50, January 17, 1985.

D. M. Pozar, "A Review of Aperture Coupled Microstrip Antennas: History, Operation, Development, and Applications", May 1996.

Dr. C. Vivek, S. Palanivel Rajan, "Z-TCAM: An Efficient Memory Architecture Based TCAM", Asian Journal of Information Technology, ISSN No.: 1682-3915, Vol. No.: 15, Issue : 3, pp. 448-454, 2016.

Egashira, S., and Nishiyama, E.1996. Stacked microstrip antenna with wide bandwidth and high gain", *IEEE Trans. Antennas Propag.*, pp. 1533–1534.

Eleftheriades, G. V., Balmain, K. G. Negative Refraction Metamaterials: Fundamental Principles and Applications. New York: IEEE Press/Wiley, 2005.

Eng Gee Lim, Zhao Wang, Jing Chen Wang, Mark Leach, Rong Zhou, Chi-Un Lei, Ka Lok Man, "Wearable Textile Substrate Patch Antenna", Engineering Letters, 22:2, EL_22_2_08

Engheta, N., Ziolkowski, R. W. Metamaterials. Physics and Engineering Explorations. New York: IEEE Press/Wiley, 2006.

F. Qureshi, M. A. Antoniades, and G. V. Eleftheriades, "A Compactand LowProfile Metamaterial Ring Antenna with Vertical polarization", *IEEE Antennas and Propagation Letters*, vol. 4, pp. 333-336, 2005.

Hemant Kumar Varshney et al., "A Survey on Different Feeding Techniques of Rectangular Microstrip Patch Antenna," *International Journal of Current Engineering and Technology*, Vol.4, No.3 (June 2014).

IE3D User Manual Release: Zeland Software, 2003.

J. G. Joshi, Shyam S. Pattnaik, and S. Devi, "Metamaterial Embedded Wearable Rectangular Microstrip Patch Antenna", International Journal of Antennas and Propagation, July 2012.

J. M. Rathod, "Comparative study of microstrip patch antenna for wireless communication application," International Journal of Innovation, Management and Technology, vol.1, no.2, pp. 194–197, 2010.

Jaume Anguera, Lluís Boada, Carles Puente, Carmen Borja and Jordi Soler. 2004. Stacked H-shaped microstrip patch antenna. IEEE Transactions on antennas and propagation, vol. 52, NO. 4.

K. R. Carver and J. W. Mink, "Microstrip antenna technology,", IEEE Transactions on Antennas and Propagation, vol. 29, no. 1, pp. 2–24, 1981.

Khan M. Z. Shams' and M. Ali, "A Capacitively coupled polymeric internal antenna," IEEE Int. Symp. On Antenna Prop. Soc. Vol 2, 20-25 June 2004, pp.1967–1970.

Krzysztofik, Wojciech & Cao, Thanh. (2019). Metamaterials in Application to Improve Antenna Parameters. 10.5772/intechopen.80636.

Lapine, M., Tretyakov, S. Contemporary notes on metamaterials. *IET Microwaves, Antennas & Propagation*, 2007, vol. 1, no. 1, p. 3 – 11.

Lavi Agarwal, Prateek Rastogi, "Design And Analysis Of Circularly Polarized Microstrip Patch Antenna Using HFSS", International Journal of Computer Applications (0975-8887), Volume 124-No. 16, August 2015

Li B., Wu B., and Liang C.-H, "Study on High Gain Circular waveguide Array antenna with Metamaterial structure", PIER 60, pp. 207–219, 2006.

Li Jiusheng, "Chamfered band based on metamaterial unit cell" *IEEE conference International Conference on Digital Object dentifier*: 10.1109/ICMMT.2007.381368

M. T. I.-u. Huque, M. Chowdhury, M. K. Hosain, and M. S. Alam,"Performance analysis of corporate feed rectangular patch element andcircular patch element 4x2 microstrip array antennas," International Journal of Advanced Computer Science and Applications, vol. 2, no. 7, 2011.

M. Z. M. Zani, M. H. Jusoh, A. A. Sulaiman, N. H. Baba, R. A. Awang, and M. F. Ain, "Circular Patch Antenna On Metamaterial", IEEE International Conference on Electronic Devices, Systems and Applications (ICEDSA), pp. 313-316, 2010.

Marqués, R., Martin, F., Sorolla, M. Metamaterials with Negative Parameters: Theory, Design and Microwave Applications. New York: Wiley-Interscience, 2008.

Odhekar, Anuja & Deshmukh, Amit (2016) "Microstrip Antenna optimization using split ring and complementary split ring resonator" 1-7. 10.1109/ICICES.2016.7518867.

P. C. Sharma, Pradeep C. Gupta," Analysis and Optimized Design of Single Feed Circularly Polarized Microstrip Antennas", IEEE Transactions on Antennas and Propagation, VolAP- 31, No. 6, November 1983.

R. Garg, Microstrip antenna design handbook. Artech House, 2001.

R. Garg, P. Bhartia, I. Bahl and A. Ittipiboon.2001. Microstrip antenna design handbook. Artech House, Boston.

R. Garg, P. Bhartia, I. Bahl, and A. Ittipiboon, "Microstrip Antenna Design Handbook" London: Artech House, 2001

R. Saluja et. al., "Analysis of Bluetooth Patch Antenna with Different Feeding Techniques using Simulation and Optimization," in *Proc. Internatinational Conf. on Microwaves*, New York, 1994, pp. 8–16.

R. A Sainati, "CAD for Microstrip Antennas for Wireless Application"', Artech House, Boston, 1996, pp. 21-63, 85-92.

R. W. Ziolkowski, 2003. "Design, fabrication, and testing of double negative metamaterials", *IEEE Trans. on Antennas and Propagation*, vol. 51, no. 7, pp. 1516-1529, July, 2003.

R. W. Ziolkowski, and A. Erentok. "Metamaterial-Based Efficient Electrically small Antennas," *IEEE Trans. on Antennas and Propagation*, Vol. 54, No. 7, pp. 2113- 2130. July, 2006

S. Palanivel Rajan, "Review and Investigations on Future Research Directions of Mobile Based Telecare System for Cardiac Surveillance", Journal of Applied Research and Technology, ISSN No.: 1665–6423, Vol. 13, Issue 4, pp. 454-460, 2015.

S. Palanivel Rajan, R. Sukanesh, "Experimental Studies on Intelligent, Wearable and Automated Wireless Mobile Tele-Alert System for Continuous Cardiac Surveillance", Journal of Applied Research and Technology, ISSN No.: 1665–6423, Vol. No. 11, Issue No.: 1, pp. 133 -143, 2013.

S. Palanivel Rajan, R. Sukanesh, "Viable Investigations and Real Time Recitation of Enhanced ECG Based Cardiac Tele-Monitoring System for Home-Care Applications: A Systematic Evaluation", Telemedicine and e-Health Journal, ISSN: 1530-5627, Vol. No.: 19, Issue No.: 4, pp. 278-286, 2013.

S. Palanivel Rajan, R. Sukanesh, S. Vijayprasath, "Analysis and Effective Implementation of Mobile Based Tele-Alert System for Enhancing Remote Health-Care Scenario", Health MED Journal, ISSN No. : 1840-2291, Vol. No. 6, Issue No. 7, pp. 2370–2377, 2012.

S. Palanivel Rajan, R. Sukanesh, S. Vijayprasath, "Design and Development of Mobile Based Smart Tele-Health Care System for Remote Patients", European Journal of Scientific Research, ISSN No.: 1450-216X/1450-202X, Vol. No. 70, Issue 1, pp. 148-158, 2012.

S. Sankaralingam, B. Gupta," Development Of Textile Antennas For Body Wearable Applications And Investigations On Their Performance On Bent conditions", Progress In Electronics Research B, Vol. 22, 53-71, 2010

Shalaev, V. M., Sarychev, A. K. Electrodynamics of Metamaterials. Singapore: World Scientific, 2007.

Shamonina, E., Solymar, L. Metamaterials: How the subject started. *Metamaterials*, 2007, vol. 1, no. 1, p. 12 – 18.

Shobhit K. Patel, Sunil Lavadiya, Y. P. Kosta, Medhavi Kosta, Truong Khang Nguyen, Vigneswaran Dhasarathan. (2020) Numerical investigation of liquid metamaterial-based superstrate microstrip radiating structure. *Physica B: Condensed Matter*, 585, pages 412095.

Solymar, L. Shamonina, E. Waves in Metamaterials. Oxford University Press, 2009.

W Wang, B.-I. Wu, J. Pacheco, X. Chen, T. Grzegorczyk and J. A. Kong, "A study of using metamaterials as antenna substrate to enhance gain", PIER 51, pp. 295–328, 2005

Yogita, Prof. Arun Shukla, "Design and Analysis of an Irregular Diamond Edge Slotted Microstrip Patch Antenna at 1.6 GHz for WLAN," *International Research Journal of Engineering and Technology (IRJET)*, Vol. 3 Issue 3, 2016

About the Authors

Professor S. Kannadhasan, PhD
Department of Electronics and Communication Engineering
Cheran College of Engineering
Tamil Nadu, India

Professor R. Nagarajan, PhD
Department of Electrical and Electronics Engineering
Gnanamani College of Technology
Tamil Nadu, India

Index

#

4G communication, v, 63, 66
5G propagation, vii

A

access, 24, 70, 72, 73, 74, 77
advanced EM material(s), vii
algorithm, 20, 40, 41
antenna array(s), vii, 26, 35, 40, 48, 66, 67
antenna measurement(s), vii
antenna theory, vii, 86
aperture antenna(s), vii
applied electromagnetics, vii
array antenna(s), vii, 20, 23, 39, 87
artificial intelligence, 19, 39

B

band gap, 16, 55, 57, 59, 65
bandwidth, 2, 3, 6, 11, 13, 14, 16, 17, 18, 19, 20, 21, 22, 24, 25, 26, 27, 28, 29, 30, 31, 33, 34, 35, 36, 37, 38, 39, 41, 43, 44, 46, 48, 49, 50, 51, 54, 55, 64, 66, 70, 78, 79, 80, 83, 85, 86
bandwidth enhancement, v, 14
Bluetooth, 23, 29, 41, 42, 44, 68, 82, 88
broadband, 29, 46, 74, 85

C

cell phone(s), 29
channel interference, 19

communication, 2, 4, 13, 20, 23, 25, 26, 32, 33, 34, 36, 37, 39, 40, 41, 42, 44, 48, 50, 51, 55, 59, 66, 69, 70, 71, 72, 73, 74, 75, 77, 80, 81, 82, 83, 87
communication system(s), 4, 20, 26, 34, 37, 39, 41, 48, 50, 70, 71, 83
communication technologies, 26, 70, 73, 81
computational electromagnetics, vii
configuration, 6, 14, 15, 19, 36, 41, 43, 44, 51, 75
connectivity, 21, 23, 24, 27, 32, 74

D

data communication, 32
data processing, 24, 37
data rate(s), 51, 66, 67, 68, 72
data transfer, 72, 73
dielectric constant, 3, 4, 32, 34, 35, 43, 44, 49, 54, 78, 79
dielectric permittivity, 55
diffraction, vii, 3, 57, 60
directional antenna(s), 67

E

electric field, 6, 7, 14, 59
electromagnetic, 8, 14, 15, 16, 21, 26, 34, 35, 45, 48, 53, 54, 55, 56, 57, 58, 59, 60, 62, 64, 65, 69
electromagnetic field(s), 8
electromagnetic wave(s), 17, 26, 34, 35, 53, 59
electronic circuit(s), 48

electronic system(s), 17
energy, 13, 23, 45, 54, 64, 66, 67, 71, 74

F

fabrication, 29, 33, 78, 88
Federal Communications Commission, 67, 77

G

global positioning system (GPS), 1, 4, 20, 25, 29, 36, 68

L

latency, 24, 25, 63, 66, 67, 73, 74
local area networks, 1, 25, 83

M

machine learning, 39
magnetic field(s), 14, 45, 54, 59, 66
manufacturing, 1, 4, 5, 15, 25, 33, 37, 51, 62, 69, 78, 82
material(s), vii, 3, 15, 43, 44, 53, 54, 55, 56, 57, 58, 59, 62, 67, 78, 79, 81, 82, 83, 84
Maxwell equations, 51
medical, vii, 4, 27, 41, 57, 62, 77, 78, 79, 82
medical microwave application(s), vii
metamaterial(s), v, vii, 53, 54, 55, 56, 57, 58, 59, 60, 61, 62, 85, 86, 87, 88, 89
metasurface(s), vii
microstrip patch antenna, v, 1, 2, 3, 4, 5, 9, 10, 11, 13, 15, 16, 17, 19, 21, 23, 25, 26, 27, 29, 30, 31, 32, 33, 34, 35, 37, 41, 42, 43, 44, 47, 48, 50, 63, 67, 70, 78, 79, 80, 85, 86, 87, 89
microwave engineering, vii
microwave radiation, 5, 53, 60
microwave(s), 51, 60

MIMO, vii, 66, 68
mobile communication, 1, 4, 25, 29, 41, 71, 75
mobile device(s), 1, 23, 66, 68, 69, 70, 71, 74
mobile phone, 24, 25, 41, 42, 44, 45, 63, 65
mobile telecommunication, 23, 69

N

nanophotonic(s), vii
navigation system, 79
network operator(s), 24, 66
networking, 1, 5, 8, 15, 18, 19, 21, 23, 25, 27, 28, 29, 35, 63, 72, 73

O

operating cost(s), 52
optical nano-antenna(s), vii
optimization, 22, 25, 40, 41, 47, 88
optimization method, 47

P

plasmonics, vii
probe, 2, 16, 19, 34, 43, 44, 49, 79

R

radar, 1, 2, 8, 14, 27, 28, 41, 45, 77, 81
radar system(s), vii, 2
radiation, 2, 3, 4, 5, 6, 7, 14, 17, 19, 20, 21, 22, 25, 26, 27, 29, 31, 33, 34, 35, 37, 38, 41, 47, 48, 49, 51, 53, 55, 60, 65, 69, 73, 78, 79, 80, 83
radio, 1, 2, 6, 7, 13, 15, 17, 27, 32, 33, 45, 50, 59, 67, 69, 71
remote sensing, 1, 2, 29

S

scattering, vii, 53
secure communication, 52
sensing, 1, 2, 29, 54
sensor node(s), 71
sensor(s), 25, 54, 55, 59, 60, 71, 78, 82
service quality, 19, 73, 74
signal quality, 72
signalling, 75
signal(s), 16, 20, 35, 37, 54, 72
signal-to-noise ratio, 36
simulation(s), 8, 17, 19, 32, 40, 41, 48, 51, 53, 67, 78, 79, 81, 83, 85
smart phone(s), 1, 65
software, 11, 19, 34, 40, 41, 49, 51, 53, 79

T

technological advancement, 11
technology, 1, 2, 5, 8, 14, 15, 17, 18, 22, 23, 24, 25, 26, 27, 28, 29, 33, 35, 37, 39, 41, 42, 44, 45, 51, 53, 62, 65, 66, 67, 68, 69, 70, 71, 72, 73, 77, 79, 80, 81, 87
telecommunication(s), 17, 23, 69
terminal(s), 15, 48, 68
transmission, 3, 7, 13, 14, 17, 20, 22, 24, 26, 29, 34, 45, 47, 53, 56, 57, 64, 70, 71, 72, 74, 81

U

universal access, 24

W

wavelength(s), 54, 57
wearable antenna(s), 77, 78, 80, 82
web, 33, 71, 73
web service, 73
Wi-Fi, 23, 29, 32, 65
wireless connectivity, 21
wireless device(s), 25, 26
wireless local area network (WLAN), 1, 21, 27, 29, 32, 36, 39, 80, 81, 83, 89
wireless network(s), 15, 19, 63, 71, 72
wireless technology, 33, 72, 80
worldwide interoperability for microwave access (WiMAX), 1, 8, 32, 36, 73